Cosmic Microwave Background Radiation

Cosmic Microwave Background Radiation

by

Trevor G. Underwood

By the same author:

"Quantum Electrodynamics – annotated sources. Volumes I and II." (April 2023);

"Special Relativity." (June 2023);

"General Relativity." (November 2023);

"Gravity." (March 2024);

"Electricity & Magnetism." (May 2024);

"Quantum Entanglement." (June 2024);

"The Standard Model." (September 2024);

"New Physics." (October 2024);

"The Cosmological Redshift of Light." (November 2024).

Published by Trevor G. Underwood
18 SE 10th Ave.
Fort Lauderdale, FL 33301

ISBN: 979-8-218-58694-2 (hardcover)

Library of Congress Control Number: 2025900129

Printed and distributed by Lulu Press, Inc.
627 Davis Dr.
Ste. 300
Morrisville, NC 27560
http://www.lulu.com/shop

Front cover: This map shows small temperature fluctuations of the *cosmic microwave background* (CMB) radiation that correspond to regions of slightly different densities, representing the seeds of the stars and galaxies of today. This map was constructed from an analysis of observations of the sky at wavelengths of light spanning 850 microns to 1 cm (353 GHz to 30 GHz). Additional observations spanning 350 to 550 microns (857 to 545 GHz) helped characterize foreground dust in the Milky Way, which was removed from the final CMB data shown here. Image courtesy of ESA/NASA.

CONTENTS.

Page no.

Owing to the *magnetic field*, these *currents* give mechanical *forces* which change the state of motion of the liquid. Thus, a kind of combined *electromagnetic-hydro-dynamic wave* is produced which, so far as I know, has as yet attracted no attention.

65   **Alfvén, H. (1954). *On the origin of the solar system"*.** Oxford, Clarendon Press. *Preface.*

66   **Erickson, W. C. (November, 1957). A Mechanism of Non-Thermal Radio-Noise Origin.** *Astrophys. J.*, 126, 480; https://adsabs.harvard.edu/pdf/1957ApJ...126..480E.
A mechanism of *non-thermal radio-noise* origin is proposed. The action of this mechanism may be summarized in the following manner. Suppose that clouds of *interstellar grains* exist in the *radio-source* regions. If a *high-velocity gas cloud* collides with a cloud of *grains*, the *grains* will be bombarded by moderately fast *atoms* and/or *ions*. These collisions will transfer *angular momentum* to the grains, and, in fact, the *angular velocity* of each grain will execute a dynamical "walk." *It is shown that rotational frequencies comparable with radio frequencies may be attained. If some of the grains possess electric or magnetic dipole moments due to polar or ferromagnetic substances or statistical fluctuations in the distribution of charge on the grains, they will radiate classically at radio frequencies.* Rather improbably high grain densities are required in order to account for the *total radio-frequency radiation of high-emissivity sources.* However, *the high-frequency portion of this radiation could be generated with moderate grain densities.*

73   **Dicke, R. H., Peebles, P. J. E., Roll, P. G., & Wilkinson, D. T. (July, 1965). Cosmic Black-body Radiation.** *Astrophys. J.*, 142, 414; https:/ /articles.adsabs.harvard.edu/pdf/ 1965ApJ...142..414D. One of the basic problems of cosmology is the *singularity characteristic* of the familiar cosmological solutions of *Einstein's field equations.* Also puzzling, is the presence of *matter* in excess over *antimatter* in the universe, for *baryons* and *leptons* are thought to be conserved. *Thus, in the framework of conventional theory we cannot understand the origin of matter or of the universe.* We can distinguish three main attempts to deal with these problems. 1. The assumption of continuous creation (Bondi & Gold, 1948; Hoyle, 1948), *which avoids the singularity by postulating a universe expanding for all time and a continuous but slow creation of new matter in the universe.* 2. The assumption (Wheeler, 1964) that *the creation of new matter is intimately related to the existence of the singularity,* and that the resolution of both paradoxes may be found in a proper quantum mechanical treatment of *Einstein's field equations.* 3. The assumption that *the singularity results from a mathematical over-idealization, the requirement of strict isotropy or uniformity,* and that it would not occur in the real world (Wheeler, 1958; Lifshitz & Khalatnikov, 1963). If this third premise is accepted tentatively as a working hypothesis, it carries with it a possible resolution of the second paradox, for the *matter* we see about us now may represent the same *baryon* content of the previous expansion of a closed universe, oscillating for all time.

82   **Arno Allan Penzias (1933–2024): A visionary explorer of the Universe.**

87   **Penzias, A. A., & Wilson, R. W. (1965). A Measurement of Excess Antenna Temperature at 4080 Mc/s.** *The Astrophysical Journal*, Letters to the Editor, 142, 1, 419–21; https://doi.org/ 10.1086/148307. Measurements of the effective *zenith noise*

temperature of the 20-foot horn-reflector antenna (Crawford, Hogg, & Hunt, 1961) at the Crawford Hill Laboratory, Holmdel, New Jersey, at 4080 Mc/s have yielded a value about 3.5° K higher than expected. This excess *temperature* is, within the limits of our observations, isotropic, unpolarized, and free from seasonal variations (July, 1964-April, 1965). A possible explanation for the observed excess *noise temperature* is the one given by Dicke, Peebles, Roll, & Wilkinson (1965) in a companion letter in this issue.

90   **Hannes Alfvén - Nobel Lecture, December 11, 1970. Plasma physics, space research and the origin of the solar system.** Hannes Alfvén – Nobel Lecture. NobelPrize.org. https://www.nobelprize.org/uploads/2018/06/alfven-lecture.pdf.

108   **Lerner, E. J. (August, 1988). Plasma model of microwave background and primordial elements: an alternative to the big bang.** *Lasers and Particle Beams*, 6, 457-69; https://doi.org/10.1017/S0263034600005395. *A plasma model of the origin of the light elements and the microwave background is presented.* In contrast to the conventional Big Bang hypothesis, *the model assumes that helium, deuterium and the microwave background were all generated by massive stars in the early stages of galaxy formation.* The *microwave background* is scattered and isotropized by multi-GeV *electrons* trapped in the jets emitted by *active galactic nuclei.* The model produces reasonable amounts of heavy elements, *accurately predicts the gamma-ray background intensity and spectrum*, and explains the statistics of *quasars, compact* and *extended radio sources.*

133 **Lerner, E. J. (October, 1992). The Big Bang Never Happened—A Reassessment of the Galactic Origin of Light Elements (GOLE) Hypothesis and its Implications.** Book (569 pages): Knopf Doubleday Publishing Group; https://www.lppfusion.com/storage/GOLE-Lerner.pdf. Far-ranging and provocative, The *Big Bang Never Happened* is more than a critique of one of the primary theories of astronomy -- that the universe appeared out of nothingness in a single cataclysmic explosion ten to twenty billion years ago. Drawing on new discoveries in particle physics and thermodynamics as well as on readings in history and philosophy, Eric J. Lerner confronts the values behind the *Big Bang theory*: the belief that mathematical formulae are superior to empirical observation; that the universe is finite and decaying; and that it could only come into being through some outside force. With inspiring boldness and scientific rigor, he offers a brilliantly orchestrated argument that generates explosive intellectual debate.

134 **Lerner, E. J. (1993). The Case Against the Big Bang.** In: Arp, H. C., Keys, C. R., Rudnicki, K. (eds) *Progress in New Cosmologies*. Springer, Boston, MA; https://doi.org/10.1007/978-1-4899-1225-1_7. Despite its widespread acceptance, the *Big Bang* theory *is presently without any observational support*. All of its quantitative predictions are contradicted by observation, and none are supported by the data. Its *predictions of light element abundances* are inconsistent with the latest data. It is impossible to produce a *Big Bang "age of the universe"* which is old enough to allow the development of the observed large-scale structures, or even the evolution of the Milky Way galaxy. The theory does not predict an isotropic *cosmic microwave background* without several additional ad hoc assumptions which are themselves clearly contradicted by observation. By contrast, *plasma cosmology* theories have provided *explanations of the light element abundances*, the *origin of large-scale structure* and the *cosmic microwave*

background that accord with observation. It is time to abandon the *Big Bang* and seek other explanations of the *Hubble relationship*.

136   **Ferrara, A., & Dettmar, R. -J. (May, 1994). Radio-emitting dust in the free electron layer of spiral galaxies: Testing the disk/halo interface.** *Astrophys. J.*, 427, 155-9; https://doi.org/10.1086/174128. We present a study of the *radio emission from rotating, charged dust grains immersed in the ionized gas constituting the thick, Hα-emitting disk of many spiral galaxies*. Using up-to-date optical constants, the *charge* on the *grains* exposed to the diffuse *galactic UV flux* has been calculated. An analytical approximation for the *grain charge* has been derived, which is then used to obtain the *grain rotation frequency*. *Grains are found to have substantial radio emission peaked at a cutoff frequency in the range 10-100~GHz*, depending on the *grain size distribution* and on the efficiency of the radiative damping of the *grain* rotation. The *dust* radio emission is compared to the *free-free emission* from the ionized gas component; some constraints on the magnetic field strength in the *observed dusty filaments* are also discussed. The model can be used to test the disk-halo interface environment in *spiral galaxies,* to determine the amount and size distribution of *dust* in their ionized component, and to investigate the rotation mechanisms for the *dust*. Numerical estimates are given for experimental purposes.

137   **Lundin, R., & Marklund, G. (January, 1995). Plasma vortex structures and the evolution of the solar system—the legacy of Hannes Alfvén.** *Physica Scripta,* 60, 198-205; https://doi.org/10.1088/0031-8949/1995/T60/023. In this report we address the present knowledge of *plasma vortex structures* and how that impacts on *cosmogony,* the *evolution of the solar system,* as it was conceived by Hannes Alfvén in 1954. Although *plasma vortex structures* are well known phenomena in contemporary *space plasma physics*, few attempts have been made to associate

plasma vortices with large astrophysical objects. Indeed, Hannes Alfvén was a pioneer in this respect by associating the *early solar system* with a *plasma vortex*. Despite obvious merits from an observational and theoretical point of view, Alfvén-cosmogony based on a synthesis between *space plasma physics* and contemporary *planetology* have had few proponents. In fact, few people outside the *space plasma physics* community appear to appreciate the evolutionary thesis by Hannes Alfvén and his coworker Gustaf Arrhenius.

138 **Lerner, E. J. (May, 1995). Intergalactic Radio Absorption and the COBE Data.** *Astrophys. Space Sci.*, 227, 1-2, 61–81; https://doi.org/10.1007/BF00678067. The COBE data on *cosmic background radiation* (CBR) *isotropy* and *spectrum* are generally considered to be explicable only in the context of the *Big Bang theory* and to be confirmation of that theory. *However, this data can also be explained by an alternative, non-Big Bang model which hypothesizes an intergalactic radio-absorbing and scattering medium.* A simple, inhomogenous model of such an *absorbing medium* can reproduce both the *isotropy* and *spectrum* of the CBR within the limits observed by COBE, and in fact *gives a better to fit to the spectrum observations than does a pure blackbody.* Such a model does not contradict any other observations, such as the existence of distant radio sources.

150 **Draine, B. T., & Lazarian, A. (February, 1998). Diffuse Galactic Emission from Spinning Dust Grains.** *Astrophys. J.*, 494, L19–L22; https://iopscience.iop.org/article/10.1086/311167/pdf. *Spinning interstellar dust grains* produce detectable rotational emission in the 10–100 GHz frequency range. *We calculate the emission spectrum and show that this emission can account for the "anomalous" galactic background component*, which correlates with *100 μm thermal emission from dust.* *Free-free emission* cannot account for the anomalous component. Implications for *cosmic background* studies are discussed.

165   **Draine, B. T., & Lazarian, A. (November, 1998). Electric Dipole Radiation from Spinning Dust Grains.** *Astrophys. J.,* 508, 157-79; https://doi.org/10.1086/306387. *We discuss the rotational excitation of small interstellar grains and the resulting electric dipole radiation from spinning dust.* Attention is given to *excitation* and *damping* of *grain rotation* by *collisions with neutrals, collisions with ions, "plasma drag", emission of infrared radiation, emission of electric dipole radiation, photoelectric emission,* and *formation of H_2 on the grain surface.* Electrostatic "focusing" can substantially enhance the rate of rotational excitation of *grains* colliding with *ions.* Under some conditions, *"plasma drag"* - due to interaction of the *electric dipole moment of the grain with the electric field produced by passing ions* – *dominates both rotational damping and rotational excitation. Emissivities* are estimated for *dust* in different phases of the *interstellar medium,* including *diffuse H_I clouds, warm H_I, low-density photo-ionized gas,* and *cold molecular gas. Spinning dust grains could explain much, and perhaps all, of the 14-50 GHz background component recently observed by Kogut, et al., de Oliveira-Costa, et al., and Leitch, et al.* Future sensitive measurements of *angular* structure in the *microwave sky brightness* from the ground and from space should detect this *emission* from *high latitude H_I clouds.* It should be possible to detect rotational emission from *small grains* by ground based pointed observations of molecular clouds, unless these *grains* are less abundant there than is currently believed.

174   **Wright, E. L. (October, 2003.) Errors in the "The Big Bang Never Happened".** https://www.astro.ucla.edu/~wright/lerner_errors.html.

        Errors in Lerner's Criticism of the Big Bang
        Errors in Lerner's Alternative to the Big Bang
        Miscellaneous Errors.

190   Lerner, E. J. (July, 2008) Dr. Wright is Wrong -- a reply to Ned Wright's "Errors in *The Big Bang Never Happened*". https://web.archive.org/web/20160108004949/http://bigbangneverhappened.org/p25.htm.

202   Fixsen, D. J. (December, 2009). The Temperature of the Cosmic Microwave Background. *Astrophys. J.*, 707, 2, 916–920; https://doi.org/10.1088/0004-637X/ 707/2/916. The *Far InfraRed Absolute Spectrophotometer* (FIRAS) data are independently recalibrated using the *Wilkinson Microwave Anisotropy Probe data* to obtain a *cosmic microwave background* (CMB) temperature of 2.7260 ± 0.0013. Measurements of the *temperature* of the CMB are reviewed. The determination from the measurements from the literature is CMB temperature of 2.72548 ± 0.00057 K.

206   Ali-Hamoud, Y. (November 12, 2012). Spinning dust radiation: a review of the theory. https://arxiv.org/pdf/ 1211.2748. This article reviews the current status of theoretical modeling of *electric dipole radiation* from *spinning dust grains*. The fundamentally simple problem of *dust grain rotation* appeals to a rich set of concepts of classical and quantum physics, owing to the diversity of processes involved. *Rotational excitation* and *damping rates* through various mechanisms are discussed, as well as methods of computing the *grain angular momentum distribution function*. Assumptions on grain properties are reviewed. The robustness of theoretical predictions now seems mostly limited by the uncertainties regarding the *grains* themselves, namely their abundance, dipole moments, size and shape distribution.

211   Kroupa, P., Pawlowski, M., & Milgrom, M. (December, 2012). The failures of the standard model of cosmology require a new paradigm. *Int. J. Mod. Phys.*, D, 21, 1230003; https://arxiv.org/pdf/1301.3907. *Cosmological models that invoke warm or cold dark matter cannot explain observed regularities in the properties of dwarf galaxies, their highly anisotropic spatial distributions, nor the correlation between observed mass discrepancies and acceleration.* These problems with the standard model of cosmology have deep implications, in particular in combination with the observation that the data are excellently described by *Modified Newtonian Dynamics (MOND)*. MOND is a classical dynamics theory *which explains the mass discrepancies in galactic systems, and in the universe at large*, without invoking 'dark' entities. MOND introduces a new universal constant of nature with the dimensions of acceleration, a_0, such that the pre-MONDian dynamics is valid for accelerations $a \gg a_0$, and the deep MONDian regime is obtained for $a \ll a_0$, where space-time scale in *variance* is invoked. Remaining challenges for MOND are (i) *explaining fully the observed mass discrepancies in galaxy clusters*, and (ii) *the development of a relativistic theory of MOND that will satisfactorily account for cosmology*. The *universal constant a_0* turns out to have an intriguing connection with cosmology: $a_0^{-} \equiv 2\pi a_0 \approx cH_0 \approx c^2(\Lambda/3)^{1/2}$. This may point to a deep connection between cosmology and internal dynamics of *local systems*.

228   **Planck's Cosmic Microwave Background (CMB) Map**

229   Schwarz D. J., Copi, C. J., Huterer, D., & Starkman, G. D. (August, 2016). CMB anomalies after Planck. *Class. Quantum Grav.*, 33, 184001; https://doi.org/10.1088/0264-9381/33/18/184001. *Several unexpected features have been observed in the microwave sky at large angular scales, both by WMAP and by Planck.* Among those features is a *lack of both variance and correlation on the largest angular scales,* *alignment of the lowest multipole moments with one another and with the motion and geometry of the solar system*, a *hemispherical power asymmetry or dipolar power modulation*, a *preference for odd parity modes* and an unexpectedly *large cold spot in the Southern hemisphere*. The individual p-values of the significance of these features are in the per mille to per cent level, when compared to the expectations of the best-fit inflationary [*relativistic*] ΛCDM model. Some pairs of those features are demonstrably uncorrelated, increasing their combined statistical significance and indicating *a significant detection of CMB features at angular scales larger than a few degrees on top of the standard model*. Despite numerous detailed investigations, we still lack a clear understanding of these large-scale features, which seem to imply a *violation of statistical isotropy and scale invariance of inflationary perturbations*. In this contribution we present a critical analysis of our current understanding and discuss several ideas of how to make further progress.

259 **Ćirković, M. M., & Perović, S. (2018). Alternative explanations of the cosmic microwave background: A historical and an epistemological perspective.** Studies in History and Philosophy of Science, Part B: Studies in History and Philosophy of Modern Physics, 62, 1–18; arXiv: 1705.07721; https://doi.org/10.1016/j.shpsb.2017.04.005. We historically trace various non-conventional explanations for the origin of the *cosmic microwave background* (CMB) and discuss their merit, while analyzing the dynamics of their rejection, as well as the relevant physical and methodological reasons for it. It turns out that there have been many such unorthodox interpretations; not only those developed in the context of theories rejecting the *relativistic* ("*Big Bang*") paradigm entirely (e.g., by Alfvén, Hoyle and Narlikar) but also those coming from the camp of original thinkers *firmly entrenched in the relativistic milieu* (e.g., by Rees, Ellis, Rowan-Robinson, Layzer and

Hively). In fact, the orthodox interpretation has only incrementally won out against the alternatives over the course of the three decades of its multi-stage development. While on the whole, none of the alternatives to the hot *Big Bang* scenario is persuasive today, we discuss the epistemic ramifications of establishing orthodoxy and eliminating alternatives in science, an issue recently discussed by philosophers and historians of science for other areas of physics. Finally, we single out some plausible and possibly fruitful ideas offered by the alternatives.

268 **Handley, W. (February, 2021). Curvature tension: evidence for a closed universe.** *Phys. Rev.*, D, 103, L041301; https://doi.org/10.1103/PhysRevD.103. L041301. The *curvature parameter tension* between *Planck* 2018, *cosmic microwave background lensing*, and *baryon acoustic oscillation data* is measured using the suspiciousness statistic to be 2.5 to 3 σ. Conclusions regarding the *spatial curvature of the universe* which stem from the combination of these data should therefore be viewed with suspicion. Without *cosmic microwave background* (CMB) *lensing* or *Baryon acoustic oscillations* (BAO), *Planck* 2018 has a moderate preference for *closed universes*, with Bayesian betting odds of over 50:1 against a *flat universe*, and over 2000:1 against an *open universe*.

277 **Abdalla, E.,** *et al.* **(June, 2022). Cosmology intertwined: A review of the particle physics, astrophysics, and cosmology associated with the cosmological tensions and anomalies.** *Journal of High Energy Astrophysics*, 34, 49-211; https://doi.org/ 10.1016/j.jheap.2022.04.002. *The standard Λ Cold Dark Matter (ΛCDM) cosmological model provides a good description of a wide range of astrophysical and cosmological data. However, there are a few big open questions that make the standard model look like an approximation to a more realistic scenario yet to be found. In this paper, we list a few important goals that need to be addressed in the next decade, taking into account the current*

discordances between the different cosmological probes, such as the *disagreement in the value of the Hubble constant* H_0, the $\sigma_8 - S_8$ *tension*, and other less statistically significant anomalies. While these discordances can still be in part the result of systematic errors, their persistence after several years of accurate analysis strongly hints at cracks in the standard cosmological scenario and the necessity for new physics or generalizations beyond the standard model. *In this paper, we focus on the tension between the Planck CMB estimate of the Hubble constant and the SH0ES collaboration measurements.* After showing the evaluations made from different teams using different methods and geometric calibrations, we list a few interesting new physics models that could alleviate this tension and discuss how the next decade's experiments will be crucial. Moreover, we focus on the tension of the *Planck* CMB data with weak lensing measurements and redshift surveys, about the value of the matter energy density Ω_m, and the amplitude or rate of the growth of structure (σ_8, $f\sigma_8$). We list a few interesting models proposed for alleviating this tension, and we discuss the importance of trying to fit a full array of data with a single model and not just one parameter at a time. Additionally, we present a wide range of other less discussed anomalies at a statistical significance level lower than the $\sigma_8 - S_8$ tensions which may also constitute hints towards *new physics*, and we discuss possible generic theoretical approaches that can collectively explain the non-standard nature of these signals. Finally, we give an overview of upgraded experiments and next-generation space missions and facilities on Earth that will be of crucial importance to address all these open questions.

279 **Lerner, E. J. (October, 2022). The Big Bang Never Happened—A Reassessment of the Galactic Origin of Light Elements (GOLE) Hypothesis and its Implications.**
https://www.lppfusion.com/storage/GOLE-Lerner.pdf. The growing list of failed predictions of the *inflationary LCDM theory* is a widely-recognized crisis in cosmology. It is timely to re-examine if the *Big Bang hypothesis* (BBH), which underlies the dominant cosmological model, is valid. The core of that hypothesis is that the universe began with a short period of extremely high temperature and density. Such a hot, dense epoch produces light elements by fusion reactions. But the published predictions of the *Big Bang Nucleosynthesis* (BBN) theory of light element production have increasingly diverged from *observations*. The predictions for both *lithium and helium abundance* now differ by many standard deviations from *observations*, a situation that is worsening at an accelerating pace. Only *deuterium* predictions have remained in agreement with *observation*. In contrast, the published predictions of *the alternative hypothesis, that all light elements were created by thermonuclear and cosmic ray processes in young galaxies, have been repeatedly confirmed by observations. This paper reassesses the galactic origin of light element (GOLE) hypothesis in light of new calculations and recent observations.* The GOLE predictions remain in good agreement with all relevant elemental abundance data sets and are contradicted by none. As well, *the expansion of space required by* BBH *is directly contradicted by both data on surface brightness and supernova light curves. Nor are any of the quantitative predictions of BBH for the CMB in accord with observations,* while the GOLE hypothesis provides an alternative explanation for the CMB that requires none of the BBH's hypothetical entities, such as *dark matter* or *dark energy*. BBH predictions are contradicted by 16 different data sets while GOLE predictions are contradicted by none. *The solution to the crisis in cosmology is to abandon the Big Bang hypothesis.*

PREFACE

The *cosmological redshift of light* is generally assumed to be due to the *Doppler effect* on light resulting from the *expansion of the universe* following the *Big Bang*. However, my previous book, [November, 2024), *The Cosmological Redshift of Light*], concluded that a better explanation is provided by Fritz Zwicky's 1929 tired-light theory as elaborated by Arthur Compton in 1923. Zwicky's *"tired-light" theory*, attributes the *linear redshift with distance from the observer* to the *loss of energy by photons*, and consequent increase in *wavelength*, resulting from *interactions between the photons and intervening electrons or matter* whilst travelling through *intergalactic* space. Compton had confirmed that scattering of electrons by photons results in a redshift due to the transfer of energy from the photons to the scattered electrons.

Cosmic microwave background (CMB) radiation is black body microwave radiation that fills all space which was first discovered in 1965 by American radio astronomers Arno Penzias and Robert Wilson. [See Penzias, A. A., & Wilson, R. W. (1965). *A Measurement of Excess Antenna Temperature at 4080 Mc/s*, below.] With a standard optical telescope, the background space between stars and galaxies is almost completely dark. However, a sufficiently sensitive radio telescope detects a faint background glow that is almost uniform and is not associated with any star, galaxy, or other object. This glow is strongest in the *microwave* region of the radio spectrum.

According to the *Big Bang theory*, which was first proposed in 1931 by Georges Lemaitre, a Roman Catholic priest, after American astronomer, Edwin Hubble reported in 1929 a linear relationship between the distance and redshift of galaxies, the *cosmic microwave background* (CMB) is *relic radiation* from the *Big Bang*. However, as measurements of the CMB improved, contradictions with this theory's predictions began to emerge and this theory began to fall apart, and alternative explanations began to emerge. The analysis in this volume suggests that the *cosmic microwave background* (CMB) radiation resulted from clouds of ionized plasma from thermonuclear reactions in stars colliding with clouds of intergalactic dust.

The first section describes the *cosmic microwave background* (CMB) *according to the Big Bang theory*. (This is based on an article in *Wikipedia* titled "Cosmic Microwave Background" which assumes *the Big Bang theory* without noting this explicitly.)

This introduction is followed by the beginnings of an alternative theory. Hannes Alfvén recognized that a *plasma* pervades the universe and applied *plasma physics* to cosmic radiation, for which he received the 1970 Nobel Prize in Physics. [Alfvén, H. (1937). *Cosmic Radiation as an Intra-galactic Phenomenon*); to magnetic storms (Alfvén, H. (1937) *A theory of magnetic storms and of the aurorae*); to electromagnetic-hydrodynamic waves (Alfvén, H. (1942). *Existence of electromagnetic-hydrodynamic waves*); and to the origin of the solar system in his book, Alfvén, H. (1954). *On the origin of the solar system*".]

> [*Plasma* is one of four fundamental states of *matter* (the other three being *solid, liquid,* and *gas*), characterized by the *presence of a significant portion of charged particles in any combination of ions or electrons*. It is the most abundant form of ordinary *matter* in the universe, mostly in *stars* (including the Sun), but also dominating the rarefied *intracluster medium* and *intergalactic medium*.]

This is followed chronologically, by the first of a series of articles describing a mechanism by which radio-frequency radiation similar to that found in the *cosmic microwave background* (CMB) can be generated by collisions between a high-velocity plasma cloud and clouds of interstellar grains. [Erickson, W. C. (November, 1957). *A Mechanism of Non-Thermal Radio-Noise Origin.*]

Penzias & Wilson (1965), noted above, showing that measurements of the effective *zenith noise temperature* yielded a value about 3.5° K higher than expected, is followed by a possible explanation in a companion letter. [Dicke, R. H., Peebles, P. J. E., Roll, P. G., & Wilkinson, D. T. (April, 1965). *Cosmic Black-body Radiation*]. This raised conflicts arising in cosmology from *Einstein's field equations* such that *neither the origin of matter nor of the universe could be understood*.

This article proposed three main attempts to deal with these problems: the assumption of continuous creation which avoids the singularity by postulating a universe expanding for all time and a continuous but slow creation of new matter in the universe; the assumption that the creation of new matter is intimately related to the existence of the singularity, and that the resolution of both paradoxes may be found in a proper quantum mechanical treatment of Einstein's field equations; and the assumption that *the singularity results from a mathematical over-idealization, the requirement of strict isotropy or uniformity*, and that it would not occur in the real world. If this third premise is accepted tentatively as a working hypothesis, it carries with it a possible resolution of the second paradox, for the *matter* we see about us now may represent the same *baryon* content [protons and neutrons] of the previous expansion of a closed universe, oscillating for all time.

In the next article Eric J. Lerner, an expert in plasma physics, presents his *plasma model of the origin of the light elements and the microwave background* as an alternative to the *Big Bang*. [Lerner, E. J. (August, 1988). *Plasma model of microwave background and primordial elements: an alternative to the big bang.*] The model assumes that helium, deuterium and the microwave background were all generated by massive stars in the early stages of galaxy formation. The microwave background is scattered and isotropized by multi-GeV electrons trapped in the *jets* emitted by active galactic nuclei. The model produces reasonable amounts of heavy elements, accurately predicts the gamma-ray background intensity and spectrum, and explains the statistics of quasars, and compact and extended radio sources.

19

Lerner's model (GOLE) was published in an expanded form in his 569-page book [Lerner, E. J. (October, 1992). *The Big Bang Never Happened—A Reassessment of the Galactic Origin of Light Elements (GOLE) Hypothesis and its Implications*], and in a chapter in Arp, H. C., Keys, C. R., Rudnicki, K. (eds) *Progress in New Cosmologies*, in which he claimed the *Big Bang theory* is presently without any observational support. By contrast, he claimed *plasma cosmology theories* have provided explanations of the *light element abundances*, the *origin of large-scale structure* and the *cosmic microwave background* that accord with observation. He noted that it was time to abandon the *Big Bang* and seek other explanations of the Hubble relationship. [See Underwood, T. G. (November, 2024), *Cosmological Redshift of Light.*]

We then return, chronologically, to a study of the radio emission from rotating, charged dust grains immersed in the ionized gas constituting the thick, Hα-emitting disk of many spiral galaxies. [Ferrara, A., & Dettmar, R. -J. (May, 1994). *Radio-emitting dust in the free electron layer of spiral galaxies: Testing the disk/halo interface.*] Grains were found to have substantial radio emission peaked at a cutoff frequency in the range 10-100~GHz, depending on the grain size distribution and on the efficiency of the radiative damping of the grain rotation.

The next article returns to *cosmic plasma physics* to address the present knowledge of plasma vortex structures and how that impacts on cosmogony, the evolution of the solar system, as it was conceived by Hannes Alfvén in 1954. [Lundin, R., & Marklund, G. (January, 1995). *Plasma vortex structures and the evolution of the solar system—the legacy of Hannes Alfvén.*]

This is followed by a response by Lerner to new COBE satellite data on *cosmic background radiation* (CBR) isotropy and spectrum which were considered to be explicable only in the context of the *Big Bang theory* and to be confirmation of that theory. Lerner claimed that the new data could also be explained by an alternative, non-Big Bang model which hypothesizes *an intergalactic radio-absorbing and scattering medium* and gives a better to fit to the spectrum observations than does a pure blackbody. [Lerner, E. J. (May, 1995). *Intergalactic Radio Absorption and the COBE Data.*]

The next two articles, by Draine and Lazarian, return to *spinning dust grains*. The first claims that spinning interstellar dust grains produce detectable rotational emission in the 10–100 GHz frequency range of which the emission spectrum can account for the "anomalous" galactic background component which correlates with 100 μm thermal emission from dust. [Draine, B. T., & Lazarian, A. (February, 1998). *Diffuse Galactic Emission from Spinning Dust Grains.*] The second discusses the rotational excitation of small interstellar grains and the resulting electric dipole radiation from spinning dust. It addresses the excitation and damping of grain rotation by collisions with neutrals, collisions with ions, "plasma drag", emission of infrared radiation, emission of electric dipole radiation, photoelectric emission, and formation of H_2 on the grain surface, claiming that spinning dust grains could explain much, and perhaps all, of the 14-50 GHz background component. [Draine, B. T., & Lazarian, A. (November, 1998). *Electric Dipole Radiation from Spinning Dust Grains.*]

This is followed in 2003 by an interesting internet article by Dr. Edward L. Wright commenting on Lerner's 1992 book "*The Big Bang Never Happened*" and Lerner's May, 1995 article *Intergalactic Radio Absorption and the COBE Data*. Wright is an American astrophysicist and cosmologist who received his PhD (Astronomy in 1976) in high-altitude rocket measurement of *cosmic microwave background* (CMB) *radiation* from Harvard University and had subsequently worked on space missions including the Cosmic Background Explorer (COBE), Wide-field Infrared Survey Explorer (WISE), and Wilkinson Microwave Anisotropy Probe (WMAP) projects. [Wright, E. L. (October, 2003.) *Errors in the "The Big Bang Never Happened".*]

21

In July 2008, Lerner responded to Wright's comments. [Lerner, E. J. (July, 2008) Dr. Wright is Wrong -- a reply to Ned Wright's "Errors in *The Big Bang Never Happened*".]

The next article reports on the latest refinement of the measurement of the *cosmic microwave background* (CMB) *temperature*, independently recalibrating the Far InfraRed Absolute Spectrophotometer (FIRAS) data using the Wilkinson Microwave Anisotropy Probe data. This confirmed the temperature of intergalactic space, to be 2.72548 ± 0.00057 K. [Fixsen, D. J. (December, 2009). *The Temperature of the Cosmic Microwave Background*.]

The next article reviews the current status of theoretical modeling of electric dipole radiation from *spinning dust grains*, concluding that robustness of theoretical predictions now seems mostly limited by the uncertainties regarding the grains themselves, namely their abundance, dipole moments, size and shape distribution. [Ali-Hamoud, Y. (November 12, 2012). *Spinning dust radiation: a review of the theory*.]

The next article notes that cosmological models that invoke warm or cold *dark matter* cannot explain observed regularities in the properties of dwarf galaxies, their highly anisotropic spatial distributions, nor the correlation between observed mass discrepancies and acceleration. It introduces an alternative in the form of *Modified Newtonian Dynamics* (MOND), a classical dynamics theory which explains the mass discrepancies in galactic systems, and in the universe at large, without invoking 'dark' entities. It concludes that the remaining challenges for MOND are (i) explaining fully the observed mass discrepancies in galaxy clusters, and (ii) the development of a relativistic theory of MOND that will satisfactorily account for cosmology. [Kroupa, P., Pawlowski, M., & Milgrom, M. (December, 2012). *The failures of the standard model of cosmology require a new paradigm*.]

This is followed by an article reporting on several unexpected features that have been observed in the microwave sky at large angular scales, both by WMAP and by the recent *Planck* satellite. Among those features was a lack of both variance and correlation on the largest angular scales, alignment of the lowest multipole moments with one another and with the motion and geometry of the solar system, a hemispherical power asymmetry or dipolar power modulation, a preference for odd parity modes and an unexpectedly large cold spot in the Southern hemisphere. It notes that despite numerous detailed investigations, we still lack a clear understanding of these large-scale features, which seem to imply a violation of statistical isotropy and scale invariance of inflationary perturbations. [Schwarz D. J., Copi, C. J., Huterer, D., & Starkman, G. D. (August, 2016). *CMB anomalies after Planck.*]

The next article, [Ćirković, M. M., & Perović, S. (2018). *Alternative explanations of the cosmic microwave background: A historical and an epistemological perspective*], traces historically various non-conventional explanations for the origin of the *cosmic microwave background* (CMB).

Will Handley [(February, 2021). (*Curvature tension: evidence for a closed universe*] notes that without *cosmic microwave background* (CMB) *lensing* or *Baryon acoustic oscillations* (BAO), *Planck* 2018 has a moderate preference for *closed universes*, with Bayesian betting odds of over 50:1 against a *flat universe*, and over 2000:1 against an *open universe*. Abdalla, E., *et al.* [(June, 2022). *Cosmology intertwined: A review of the particle physics, astrophysics, and cosmology associated with the cosmological tensions and anomalies*] notes that the standard Λ Cold Dark Matter (ΛCDM) cosmological model provides a good description of a wide range of astrophysical and cosmological data, but there are a few big open questions that make the standard model look like an approximation to a more realistic scenario yet to be found.

Finally, the most recent online update by Lerner, E. J. [(October, 2022). *The Big Bang Never Happened—A Reassessment of the Galactic Origin of Light Elements (GOLE) Hypothesis and its Implications*] notes that the growing list of failed predictions of the *inflationary LCDM theory* is a widely-recognized crisis in cosmology; it is therefore timely to re-examine if the *Big Bang hypothesis* (BBH), which underlies the dominant cosmological model, is valid. The core of that hypothesis is that the universe began with a short period of extremely high temperature and density. Such a hot, dense epoch produces light elements by fusion reactions. But the actual published predictions of the *Big Bang Nucleosynthesis* (BBN) theory of light element production have increasingly diverged from *observations*. The predictions for both *lithium and helium abundance* now differ by many standard deviations from *observations*, a situation that is worsening at an accelerating pace. Only *deuterium* predictions have remained in agreement with *observation*. In contrast, the published predictions of *the alternative hypothesis, that all light elements were created by thermonuclear and cosmic ray processes in young galaxies, have been repeatedly confirmed by observations. This paper reassesses the galactic origin of light element (GOLE) hypothesis in light of new calculations and recent observations.* The GOLE predictions remain in good agreement with all relevant elemental abundance data sets and are contradicted by none. *Only a single new assumption of EM energy loss with distance is required.*

> [This is now provided by Zwicky's tired-light theory as elaborated by Compton. See Underwood, T. G. (November, 2024). *Cosmological Redshift of Light*.]

Nor are any of the quantitative predictions of BBH for the CMB in accord with observations, while *Galactic Origin of Light Elements* (GOLE) hypothesis provides an alternative explanation for the CMB that requires none of the BBH's hypothetical entities, such as *dark matter* or *dark energy*. BBH predictions are contradicted by 16 different data sets while GOLE predictions are contradicted by none. *The solution to the crisis in cosmology is to abandon the Big Bang hypothesis.*

The current volume concludes that *there was no Big Bang, the universe is not just 13.8 billion years old*, but *is indefinitely old, and in a steady state, not expanding.*

Cosmic microwave background (CMB) according to the Big Bang theory.

[This is based on an article in *Wikipedia* with the title "Cosmic microwave background", *which assumed the Big Bang theory*.]

The *cosmic microwave background* (CMB) is microwave radiation that fills all space in the observable universe. With a standard optical telescope, the background space between stars and galaxies is almost completely dark. However, a sufficiently sensitive radio telescope detects a faint background glow that is almost uniform and is not associated with any star, galaxy, or other object. This glow is strongest in the *microwave* region of the radio spectrum.

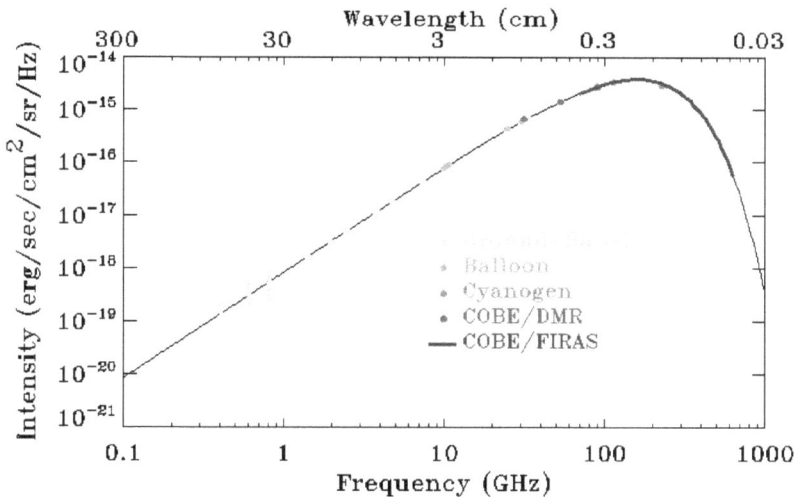

Spectral intensity is the *radiant flux* emitted, reflected, transmitted, or received *per unit solid angle* and *per unit frequency* or wavelength.

[This graph *redrawn against a linear frequency scale* appears as an almost horizontal block at an *intensity* rising sharply from 10^{-20} to 10^{-15} erg^{-1} cm^{-2} sr^{-1} at around 10 GHz, then rising slowly to a peak of about 5×10^{-15} erg^{-1} cm^{-2} sr^{-1} at around 110 GHz, then falling back at around 630 GHz, where sr is a steradian, which is the *unit of solid angle* measure (there are 4π steradians in a complete sphere).]

The *frequency* of *cosmic background* (electromagnetic) *radiation* (CBR) ranges from about 0.3 GHz to about 1000 GHz.

[The radio frequencies from 0.3 GHz to 300 GHz are referred to as *microwaves* (wavelength from about one meter to one millimeter). *Microwaves* travel by line-of-sight; unlike lower frequency radio waves, they do not diffract around hills, follow the earth's surface as ground waves, or reflect from the ionosphere, so terrestrial *microwave* communication links are limited by the visual horizon to about 64 km. At the high end of the band, they are absorbed by gases in the atmosphere, limiting practical communication distances to around a kilometer. The frequencies above 300 GHz (wavelengths above one meter) are in the *far-infrared* (which ranges from 300 GHz to 30 THz) The lower part of this range may also be called microwaves or terahertz waves. This radiation is typically absorbed by so-called rotational modes in gas-phase molecules, by molecular motions in liquids, and by phonons in solids.]

The *cosmic microwave background* (CMB) has been used as landmark evidence of the *Big Bang theory* for the origin of the universe. In the *Big Bang* cosmological models, during the earliest periods, the universe was filled with an opaque fog of dense, hot *plasma* of sub-atomic particles.

[*Plasma* is one of four fundamental states of *matter* (the other three being *solid*, *liquid*, and *gas*) characterized by the *presence of a significant portion of charged particles in any combination of ions or electrons*. It is the most abundant form of ordinary *matter* in the universe, mostly in *stars* (including the Sun), but also dominating the rarefied *intracluster medium* and *intergalactic medium*.]

[According to the *Big Bang theory*,] as the universe expanded, this *plasma* cooled to the point where *protons* and *electrons* combined to form neutral atoms of mostly *hydrogen*. Unlike the *plasma*, these atoms could not scatter thermal radiation by *Thomson scattering*, and so the universe became transparent. Known as the *recombination epoch*, this *decoupling* event released *photons* to travel freely through space. However, *the photons have grown less energetic due to the cosmological redshift associated with the expansion of the universe.*

> [This is a reversal of the previous argument that *the cosmological redshift was due to the expansion of space*, not due to the *photons* losing energy as they travel through intergalactic space.]

The *surface of last scattering* refers to a shell at the right distance in space so photons are now received that were originally emitted at the time of decoupling.

The *cosmic microwave background* (CMB) is not completely smooth and uniform, showing a faint *anisotropy* that can be mapped by sensitive detectors.

> [*Anisotropy* is the structural property of non-uniformity in different directions. An anisotropic object or pattern has properties that differ according to direction of measurement.]

Ground and space-based experiments such as COBE, WMAP and *Planck* have been used to measure these temperature inhomogeneities. [According to the *Big Bang theory*,] the *anisotropy* structure is determined by various interactions of *matter* and *photons* up to the point of *decoupling*, which results in a characteristic lumpy pattern that varies with angular scale. The distribution of the *anisotropy* across the sky has

27

frequency components that can be represented by a power spectrum displaying a sequence of peaks and valleys.

[According to the *Big Bang theory*,] the peak values of this spectrum hold important information about the physical properties of the early universe: the first peak determines the overall *curvature* of the universe, while the second and third peak detail the *density* of *normal matter* and so-called *dark matter*, respectively.

Extracting fine details from the *cosmic microwave background* (CMB) data can be challenging, since the emission has undergone modification by foreground features such as *galaxy clusters*.

[*Star clusters* are large groups of *stars* held together by *self-gravitation*. Two main types of *star clusters* can be distinguished. *Globular clusters* are tight groups of ten thousand to millions of old stars which are *gravitationally bound*. *Open clusters* are more loosely clustered groups of *stars*, generally containing fewer than a few hundred members, that are often very young. As they move through the *galaxy*, over time, *open clusters* become disrupted by the gravitational influence of *giant molecular clouds*. Even though they are no longer *gravitationally bound*, they will continue to move in broadly the same direction through space and are then known as *stellar associations*, sometimes referred to as moving groups.

Establishing precise distances to *open clusters* enables the calibration of the *period-luminosity* relationship shown by *Cepheids* variable stars, which are then used as standard candles. *Cepheids* are *luminous* and can be used to establish both the distances to remote *galaxies* and the *Hubble constant*. The *open cluster* NGC 7790 hosts three classical *Cepheids* which are critical for such efforts. [Sandage, A. (1958). Cepheids in Galactic Clusters. I. CF Cass in NGC 7790. *Astrophys. J.*, 128, 150; https://doi.org/10.1086/146532.]]

The *cosmic microwave background* radiation is an emission of uniform *black body thermal energy coming from all directions*. The *intensity* of the CMB is expressed in kelvin (K), the SI unit of temperature. The CMB has a *thermal black body spectrum* at a *temperature* of 2.72548 ± 0.00057 K. [4]

> [4] Fixsen, D. J. (2009). The Temperature of the Cosmic Microwave Background. *Astrophys. J.*, 707, 2, 916–920; arXiv:0911.1955; https://doi.org/10.1088/0004-637X/707/2/916. See below.

Variations in *intensity* are expressed as variations in *temperature*. The *blackbody temperature* uniquely characterizes the *intensity* of the radiation at all wavelengths; a measured *brightness temperature* at any wavelength can be converted to a *blackbody temperature*[5].

> [5] Wright, E. *Cosmic Microwave Background*. astro.ucla.edu.

The radiation is remarkably uniform across the sky, very unlike the almost point-like structure of *stars* or clumps of stars in *galaxies*[6].

> [6] Hu, W., & Dodelson, S. (September, 2002). Cosmic Microwave Background Anisotropies. *Annual Review of Astronomy and Astrophysics*, 40, 1, 171–216; arXiv:astro-ph/0110414; https://doi.org/10.1146/annurev.astro.40.060401.093926.

The radiation is *isotropic* to roughly one part in 25,000: the root mean square variations are just over 100 μK,[7] *after subtracting a dipole anisotropy from the Doppler shift of the background radiation.*

> [7] The Planck Collaboration (2020). Planck 2018 results V. CMB power spectra and likelihoods. *Astronomy and Astrophysics*, 641, A5; arXiv:1907.12875; https://doi.org/10.1051/0004-6361/201936386.

The latter is caused by the *peculiar velocity* of the Sun relative to the *comoving cosmic rest frame* as it moves at 369.82 ± 0.11 km/s towards the constellation *Crater* near its boundary with the constellation *Leo*[8].

[8] The Planck Collaboration (2020). Planck 2018 results. I. Overview, and the cosmological legacy of Planck. *Astronomy and Astrophysics*, 641, A1; arXiv:1807.06205; https://doi.org/10.1051/0004-6361/201833880.

The CMB *dipole* and aberration at higher multipoles have been measured, consistent with galactic motion[9].

[9] The Planck Collaboration (2014). Planck 2013 results. XXVII. Doppler boosting of the CMB: Eppur si muove. *Astronomy*, 571, 27, A27; arXiv:1303.5087; https://doi.org/10.1051/0004-6361/201321556.

Despite the very small degree of anisotropy in the *cosmic microwave background* (CMB), many aspects can be measured with high precision and such measurements are critical for cosmological theories[6].

In addition to *temperature anisotropy*, the CMB *should have* an angular variation in *polarization*. The *polarization* at each direction in the sky has an orientation described in terms of E-mode and B-mode polarization. The E-mode signal is a factor of 10 less strong than the temperature anisotropy; it supplements the temperature data as they are correlated. The B-mode signal is even weaker but may contain additional cosmological data[6].

The *anisotropy* is related to physical origin of the *polarization*. Excitation of an electron by linear polarized light generates polarized light at 90 degrees to the incident direction. If the incoming radiation is isotropic, different incoming directions create polarizations that cancel out. If the incoming radiation has quadrupole anisotropy, residual polarization will be seen[10].

[10] Hu, W. & White, M. (1997). *A CMB polarization primer*. arXiv preprint, astro-ph/9706147.

Other than the *temperature* and *polarization anisotropy*, the CMB *frequency spectrum* is expected to feature tiny departures from the *black-body law* known as spectral distortions.

These are also at the focus of an active research effort with the hope of a first measurement within the forthcoming decades, as [*according to the Big Bang theory*,] they contain a wealth of information about the primordial universe and the formation of structures at late time[11].

[11] Chluba, J., *et al.* (2021). New Horizons in Cosmology with Spectral Distortions of the Cosmic Microwave Background. *Voyage 2050 Proposals*, 51, 3, 1515–54; arXiv:1909.01593; https://doi.org/10.1007/s10686-021-09729-5.

The *cosmic microwave background* (CMB) contains the vast majority of *photons* in the universe by a factor of 400 to 1;[12] the *number density* of *photons* in the CMB is one billion times (10^9) the *number density* of *matter* in the universe.

[12] Ćirković, M. M., & Perović, S. (2018). Alternative explanations of the cosmic microwave background: A historical and an epistemological perspective. *Studies in History and Philosophy of Science*, Part B: Studies in History and Philosophy of Modern Physics, 62, 1–18; arXiv:1705.07721; https://doi.org/10.1016/j.shpsb.2017.04.005.

[*According to the Big Bang theory*,] without the *expansion of the universe* to cause the cooling of the *cosmic microwave background* (CMB), the night sky would shine as brightly as the Sun[13].

[13] Olive, K.A., & Peacock, J.A. (September, 2017). *21. Big-Bang Cosmology* in S. Navas *et al.* (Particle Data Group), to be published in (2024). *Phys. Rev.*, D 110, 030001.

The *energy density* of the *cosmic microwave background* (CMB) is 0.260 eV/cm3 (4.17×10^{-14} J/m^3), about 411 *photons*/cm^3.

[14] 29. *Cosmic Microwave Background*: Particle Data Group. Zyla, P.A., *et al*. LBL, Berkeley. (PDF).

History

When Georges Lemaître speculated in 1931 that remnants of the early universe may be observable as radiation, his candidate was *cosmic rays*[15].

[15] Peebles, P. J. E (1993). *Principles of Physical Cosmology*. Princeton University Press, pp. 139–48.

[*In support of the Big Bang theory*,] Richard C. Tolman showed in 1934 that *expansion of the universe* would cool *blackbody radiation* while maintaining a thermal spectrum.

The *cosmic microwave background* was first predicted in 1948 by Ralph Alpher and Robert Herman, in a correction[16] they prepared for a paper by Alpher's PhD advisor George Gamow[17].

[16] Alpher, R. A., & Herman, R. C. (1948). Evolution of the Universe. *Nature*, 162, 4124, 774–5; https://doi.org/10.1038/162774b0.
[17] Gamow, G. (1948). The evolution of the universe. *Nature*, 162, 4122, 680–2; https://doi.org/10.1038/162680a0.

Alpher and Herman reasoned that *if there was a Big Bang*, the *expansion of the universe* would have stretched the *high-energy radiation* of the very early universe into the *microwave region* of the *electromagnetic spectrum*. They predicted the *cosmic microwave background* (CMB) and were able to estimate the *temperature* of the *cosmic microwave background* to be 5 K[18].

[18] Assis, A. K. T., & Neves, M. C. D. (1995). History of the 2.7 K Temperature Prior to Penzias and Wilson. (PDF). *Apeiron*, 3, 79–87.

[*According to the Big Bang theory*,] they were slightly off with their estimate, but they had the *right idea*. It took another 15 years for Penzias and Wilson to discover that the *microwave background* was actually there[18].

The first published recognition of the *cosmic microwave background* (CMB) radiation as a detectable phenomenon appeared in a brief paper by Soviet astrophysicists A. G. Doroshkevich and Igor Novikov, in the spring of 1964[20].

[20] Penzias, A. A. (2006). The origin of elements (PDF). *Science*, 205, 4406, 549–54; https://doi.org/10.1126/science.205.4406.549.

In 1964, David Todd Wilkinson and Peter Roll, Dicke's colleagues at Princeton University, began constructing a Dicke radiometer to measure the *cosmic microwave background*[21].

[21] Dicke, R. H. (1946). The Measurement of Thermal Radiation at Microwave Frequencies. *Review of Scientific Instruments*, 17, 7, 268–75; https://doi.org/10.1063/1.1770483. This basic design for a radiometer has been used in most subsequent *cosmic microwave background* experiments.

In 1964, Arno Penzias and Robert Woodrow Wilson at the Crawford Hill location of Bell Telephone Laboratories in nearby Holmdel Township, New Jersey had built a Dicke radiometer that they intended to use for radio astronomy and satellite communication experiments. The antenna was constructed in 1959 to support Project Echo—the National Aeronautics and Space Administration's passive communications satellites, which used large earth orbiting aluminized plastic balloons as reflectors to bounce radio signals from one point on the Earth to another[19].

[19] Overbye, D. (September 5, 2023). Back to New Jersey, Where the Universe Began - A half-century ago, a radio telescope in Holmdel, N.J., sent two astronomers 13.8 billion years back in time — and opened a cosmic window that scientists have been peering through ever since. *The New York Times*.

On 20 May 1964 they made their first measurement clearly showing the presence of the *microwave background*[22], with their instrument having an excess 4.2K antenna temperature which they could not account for.

[22] *The Cosmic Microwave Background Radiation* (Nobel Lecture) by Robert Wilson (December 8, 1978), p. 474.

After receiving a telephone call from Crawford Hill, Dicke said "Boys, we've been scooped[23,24,25]."

[23] Penzias, A. A., & Wilson, R. W. (1965). A Measurement of Excess Antenna Temperature at 4080 Mc/s. *The Astrophysical Journal*, 142, 1, 419–21; https://doi.org/10.1086/148307.
[24] Smoot Group (March 28, 1996). *The Cosmic Microwave Background Radiation*. Lawrence Berkeley Lab. Retrieved 2008-12-11.
[25] Dicke, R. H., *et al*. (1965). Cosmic Black-Body Radiation. *Astrophysical Journal*, 142, 414–9; https://doi.org/10.1086/148306.

A meeting between the Princeton and Crawford Hill groups determined that the antenna *temperature* was indeed due to the *microwave background*. Penzias and Wilson received the 1978 Nobel Prize in Physics for their discovery[26].

[26] (1978). *The Nobel Prize in Physics 1978*. Nobel Foundation.

The interpretation of the *cosmic microwave background* was a controversial issue in the late 1960s. Alternative explanations included *energy* from *within the solar system*, from *galaxies*, from *intergalactic plasma*, from *multiple extragalactic radio sources*. Two requirements would show that the *microwave radiation* was truly "*cosmic*". First the *intensity vs frequency* or spectrum needed to be shown to match a thermal or blackbody source. This was accomplished by 1968 in a series of measurements of the radiation *temperature* at higher and lower *wavelengths*. Second the radiation needed be shown to be *isotropic*, the same from all directions. This was also accomplished by 1970, demonstrating that this radiation was truly cosmic in origin[27].

[27] Partridge, R. B. (2019-04-04). *The cosmic microwave background: from discovery to precision cosmology*. In Kragh, Helge; Longair, Malcolm S. (eds.). *The Oxford Handbook of the History of Modern Cosmology* (1st ed.). Oxford University Press. pp. 292–345. https://doi.org/10.1093/oxfordhb/9780198817666.013.8.

In the 1970s numerous studies [*in support of the Big Bang theory*,] showed that tiny deviations from isotropy in the *cosmic microwave background* (CMB) could result from events in the *early universe*[27]. Harrison[28], Peebles and Yu[29], and Zel'dovich[30] realized that the *early universe* would require quantum inhomogeneities that would result in temperature *anisotropy* at the level of 10^{-4} or 10^{-5}. [27]

[28] Harrison, E. R. (1970). Fluctuations at the threshold of classical cosmology. *Phys. Rev.*, D, 1, 10, 2726–30; https://doi.org/10.1103/PhysRevD.1.2726.

[29] Peebles, P. J. E., & Yu, J. T. (1970). Primeval Adiabatic Perturbation in an Expanding Universe. *Astrophys. J.*, 162, 815–36; https://doi.org/10.1086/150713.

[30] Zeldovich, Y. B. (1972). A hypothesis, unifying the structure and the entropy of the Universe. *MNRAS*, 160, 1P–4P; https://doi.org/10.1093/mnras/160.1.1P.

Rashid Sunyaev, using the alternative name *relic radiation*, calculated the observable imprint that these inhomogeneities would have on the *cosmic microwave background*[31].

[31] Sunyaev, R. A., & Zel'dovich, Y. B. (1970). Small-scale fluctuations of relic radiation. *Astrophys. Space Sci.*, 7, 1, 3–19; https://doi.org/10.1007/BF00653471.

After a lull in the 1970s caused in part by the *many experimental difficulties* in measuring *cosmic microwave background* (CMB) at high precision[27], increasingly stringent limits on the *anisotropy* of the *cosmic microwave background* were set by ground-based experiments during the 1980s. RELIKT-1, a Soviet *cosmic microwave background anisotropy* experiment on board the Prognoz 9 satellite (launched 1 July 1983), gave the first upper limits on the large-scale *anisotropy*[27].

[The other key event in the 1980s was the proposal by Alan Guth for *cosmic inflation*. This theory (*based on the Big Bang theory*) of *rapid spatial expansion* gave an explanation for large-scale isotropy by allowing causal connection just before the epoch of *last scattering*[27]. With this and similar theories, detailed prediction encouraged larger and more ambitious experiments.]

The NASA Cosmic Background Explorer (COBE) satellite orbited Earth in 1989–1996 detected and quantified the large-scale *anisotropies* at the limit of its detection capabilities. The NASA COBE mission clearly confirmed the primary *anisotropy* with the Differential Microwave Radiometer instrument, publishing their findings in 1992[32,33].

[32] Smoot, G. F., *et al.* (1992). Structure in the COBE differential microwave radiometer first-year maps. *Astrophysical Journal Letters*, 396, 1, L1–L5; https://doi.org/10.1086/ 186504.

[33] Bennett, C. L., *et al.* (1996). Four-Year COBE DMR Cosmic Microwave Background Observations: Maps and Basic Results. *Astrophysical Journal Letters*, 464, L1–L4; arXiv:astro-ph/9601067; https://doi.org/10.1086/310075.

The Nobel Prize in Physics 2006 was awarded jointly to Mather and Smoot for their discovery of the blackbody form and anisotropy of the cosmic microwave background radiation.

Inspired by the COBE results, a series of ground and balloon-based experiments measured *cosmic microwave background anisotropies* on smaller angular scales over the two decades. The sensitivity of the new experiments improved dramatically, with a reduction in internal noise by three orders of magnitude[1].

[1] Komatsu, E. (May, 2022). New physics from the polarized light of the cosmic microwave background. *Nature Reviews Physics*, 4, 7, 452–69; arXiv:2202.13919; https://doi.org/10.1038/s42254-022-00452-4.

The primary goal of these experiments was to measure the scale of the first *acoustic peak*, which COBE did not have sufficient resolution to resolve.

[*According to the Big Bang theory*,] this peak corresponds to large scale density variations in the early universe that are created by gravitational instabilities, resulting in acoustical oscillations in the *plasma*[34].

[34] Grupen, C., *et al.* (2005). *Astroparticle Physics*. Springer, pp. 240–241.

During the 1990s, the first peak in the *anisotropy* was measured with increasing sensitivity and was tentatively detected by the MAT/TOCO experiment[35]. By 2000 the result was confirmed by the BOOMERanG[36] and MAXIMA experiments[37], which reported that the highest power fluctuations occur at scales of approximately one degree.

[35] Miller, A. D., *et al.* (1999). A Measurement of the Angular Power Spectrum of the Microwave Background Made from the High Chilean Andes. *Astrophys. J.*, 521, 2, L79–L82; arXiv:astro-ph/9905100; https://doi.org/10.1086/312197.

[36] Melchiorri, A., *et al.* (2000). A Measurement of Ω from the North American Test Flight of Boomerang. *The Astrophysical Journal Letters*, 536, 2, L63–L66; arXiv:astro-ph/9911445; https://doi.org/10.1086/312744.

[37] Hanany, S., *et al.* (2000). MAXIMA-1: A Measurement of the Cosmic Microwave Background Anisotropy on Angular Scales of 10'–5°. *Astrophys. J.*, 545, 1, L5–L9; arXiv:astro-ph/0005123; https://doi.org/10.1086/317322.

[*According to the Big Bang theory*,] these measurements demonstrated that the geometry of the universe is approximately flat, rather than curved[38]. They ruled out *cosmic strings* as the leading theory of cosmic structure formation and suggested *cosmic inflation* was the right theory of structure formation[39].

[38] de Bernardis, P., *et al.* (2000). A flat Universe from high-resolution maps of the cosmic microwave background radiation. *Nature*, 404, 6781, 955–9; arXiv:astro-ph/0004404;https://doi.org/10.1038/35010035.

[39] Pogosian, L., *et al.* (2003). Observational constraints on cosmic string production during brane inflation. *Phys. Rev.*, D, 68, 2, 023506; arXiv:hep-th/0304188;https://doi.org/10.1103/ PhysRevD.68.023506.

A number of ground-based *interferometers* provided measurements of the fluctuations with higher accuracy over the next three years, including the Very Small Array, Degree Angular Scale Interferometer (DASI), and the *Cosmic Background Imager* (CBI). DASI made the first detection of the *polarization* of the *cosmic microwave background* (CMB) and the CBI provided the first *E-mode polarization spectrum* with compelling evidence that it is out of phase with the *T-mode spectrum.*

…

In June 2001, NASA launched a second CMB space mission, WMAP, to make much more precise measurements of the large-scale *anisotropies* over the full sky. WMAP used symmetric, rapid-multi-modulated scanning, rapid switching radiometers at five frequencies to minimize non-sky signal noise[40].

[40] Bennett, C. L., *et al.* (WMAP collaboration) (2003). First-year Wilkinson Microwave Anisotropy Probe (WMAP) observations: preliminary maps and basic results. *Astrophys. J., Supplement Series*, 148, 1, 1–27; arXiv:astro-ph/0302207; https://doi.org/10.1086/377253. This paper warns that "the statistics of this internal linear combination map are complex and inappropriate for most *cosmic microwave background* (CMB) analyses."

The data from the mission was released in five installments, the last being the nine-year summary.

[*According to the Big Bang theory,*] the results are *broadly consistent* with *Lambda CDM models* based on 6 free parameters and fitting in to *Big Bang cosmology* with cosmic inflation[41].

> [41] Bennett, C. L., *et al*. (September, 2013). NINE-YEAR WILKINSON MICROWAVE ANISOTROPY PROBE (WMAP) OBSERVATIONS: FINAL MAPS AND RESULTS. *Astrophys. J. Supplement Series*, 208, 2, 20; arXiv:1212.5225; https://doi.org/10.1088/0067-0049/208/2/20.

The Degree Angular Scale Interferometer (DASI) was a telescope installed at the U.S. National Science Foundation's Amundsen–Scott South Pole Station in Antarctica. It was a 13-element interferometer operating between 26 and 36 GHz (Ka band) in ten bands. The instrument is similar in design to the Cosmic Background Imager (CBI) and the Very Small Array (VSA).

In 2001 The DASI team announced the most detailed measurements of the *temperature*, or *power spectrum* of the *cosmic microwave background* (CMB).

These results contained the first detection of the 2nd and 3rd acoustic peaks in the *cosmic microwave background* (CMB), which, [*according to the Big Bang theory,*] were important evidence for *inflation theory*. This announcement was done in conjunction with the BOOMERanG and MAXIMA experiment[42].

> [42] Glanz, J. (April 30, 2001). Listen Closely: From Tiny Hum Came Big Bang. *The New York Times*.

In 2002 the team reported the first detection of *polarization anisotropies* in the CMB[43].

[43] Leitch, E.M., et al. (December, 2002). Measurement of polarization with the Degree Angular Scale Interferometer. *Nature*, 420, 6917, 763–71; arXiv:astro-ph/0209476; https://doi.org/10.1038/nature01271.

The *cosmic microwave background* (CMB) is polarized at the level of a few microkelvin. There are two types of polarization, called *E-mode* (or gradient-mode) and *B-mode* (or curl mode)[74].

[74] Trippe, S. (2014). Polarization and Polarimetry: A Review. *Journal of the Korean Astronomical Society*, 47, 1, 15–39; arXiv:1401.1911; https://doi.org/10.5303/JKAS.2014.47.1.15.

This is in analogy to electrostatics, in which the *electric field* (E-field) has a vanishing curl and the *magnetic field* (B-field) has a vanishing divergence.

The *E-modes* arise from *Thomson scattering* in a heterogeneous plasma[74]. *E-modes* were first seen in 2002 by the *Degree Angular Scale Interferometer* (DASI)[75,76].

[75] Kovac, J. M., Leitch, E. M., Pryke, C., Carlstrom, J. E., Halverson, N. W., & Holzapfel, W. L. (December, 2002). Detection of polarization in the cosmic microwave background using DASI. *Nature*, 420, 6917, 772–87; arXiv:astro-ph/0209478; https://doi.org/10.1038/nature01269.

[76] Ade, P. A. R., Aikin, R. W., Barkats, D., Benton, S. J., Bischoff, C. A., et al.. (June, 2014). Detection of B -Mode Polarization at Degree Angular Scales by BICEP2. *Phys. Rev. Lett.*, 112, 24, 241101; arXiv:1403.3985; https://doi.org/10.1103/PhysRevLett. 112.241101.

B-modes are expected to be an order of magnitude weaker than the *E-modes*.

[*According to the Big Bang theory,*] the former are not produced by standard scalar type perturbations, but are generated by *gravitational waves* during *cosmic inflation* shortly after the *Big Bang*[77,78,79].

[77] Seljak, U. (June, 1997). Measuring Polarization in the Cosmic Microwave Background. *Astrophys. J.*, 482, 1, 6–16; arXiv:astro-ph/9608131; https://doi.org/10.1086/304123.

[78] Seljak, U., & Zaldarriaga, M. (March, 1997). Signature of Gravity Waves in the Polarization of the Microwave Background. *Phys. Rev. Lett.*, 78, 11, 2054–7; arXiv:astro-ph/9609169; https://doi.org/10.1103/PhysRevLett.78.2054.

[79] Kamionkowski, M., Kosowsky, A., & Stebbins, A. (1997). A Probe of Primordial Gravity Waves and Vorticity. *Phys. Rev. Lett.*, 78, 11, 2058–61; arXiv:astro-ph/9609132;https://doi.org/10.1103/PhysRevLett.78.2058.

However, *gravitational lensing* of the stronger *E-modes* can also produce *B-mode* polarization[80,81].

[80] Zaldarriaga, M., & Seljak U. (July, 1998). Gravitational lensing effect on cosmic microwave background polarization. *Phys. Rev.*, D2, 58, 2, 023003; arXiv:astro-ph/9803150; https://doi.org/10.1103/PhysRevD.58.023003.

[81] Lewis, A., & Challinor, A. (2006). Weak gravitational lensing of the CMB. *Physics Reports*, 429, 1, 1–65; arXiv:astro-ph/0601594; https://doi.org/10.1016/j.physrep. 2006.03.002.

Detecting the original *B-modes* signal requires analysis of the contamination caused by lensing of the relatively strong *E-mode* signal[82].

[82] Hanson, D., *et al*. (2013). Detection of B-mode polarization in the Cosmic Microwave Background with data from the South Pole Telescope. *Phys. Rev. Lett.*, 111, 14, 141301; arXiv:1307.5830; https://doi.org/10.1103/PhysRevLett.111.141301.

Models of "slow-roll" *cosmic inflation* in the early universe predict primordial *gravitational waves* that would impact the polarization of the *cosmic microwave* background, creating a specific pattern of *B-mode* polarization. *Detection of this pattern would support the theory of inflation* and their strength can confirm and exclude different models of inflation[78,83].

[83] Kamionkowski, M., & Kovetz, E. D. (September, 2016). The Quest for B Modes from Inflationary Gravitational Waves. *Annual Review of Astronomy and Astrophysics*, 54, 1, 227–69; arXiv:1510.06042; https://doi.org/10.1146/annurev-astro-081915-023433.

Claims that this characteristic pattern of B-mode polarization had been measured by BICEP2 instrument[84] *were later attributed to cosmic dust* due to new results of the *Planck* experiment[85,83].

[84] Overbye, D. (September 22, 2014). Study Confirms Criticism of Big Bang Finding. *The New York Times*.
[85] Planck Collaboration Team (February, 2016). Planck intermediate results. XXX. The angular power spectrum of polarized dust emission at intermediate and high Galactic latitudes. *Astronomy & Astrophysics*, 586, 133, A133; arXiv:1409.5738; https://doi.org/10.1051/0004-6361/201425034.

The second type of B-modes was discovered in 2013 using the South Pole Telescope with help from the Herschel Space Observatory[86].

[86] Reich, E. S. (2013). Polarization detected in Big Bang's echo. *Nature*, https://doi.org/10.1038/nature.2013.13441.

In October 2014, a measurement of the B-mode polarization at 150 GHz was published by the POLARBEAR experiment[87].

[87] The Polarbear Collaboration (2014). A Measurement of the Cosmic Microwave Background B-Mode Polarization Power Spectrum at Sub-Degree Scales with POLARBEAR. *Astrophys. J.*, 794, 2, 171; arXiv:1403.2369; https://doi.org/10.1088/0004-637X/794/2/171.

Compared to BICEP2, POLARBEAR focuses on a smaller patch of the sky and is less susceptible to dust effects. The team reported that POLARBEAR's measured B-mode *polarization was of cosmological origin* (and not just due to dust) at a 97.2% confidence level.

...

A third space mission, the ESA (European Space Agency) *Planck Surveyor*, was launched in May 2009 and performed an even more detailed investigation until *it was shut down in October 2013*. *Planck* employed both HEMT radiometers and bolometer technology and measured the CMB at a smaller scale than WMAP. Its detectors were trialed in the Antarctic Viper telescope as ACBAR (Arcminute Cosmology Bolometer Array Receiver) experiment—which has produced the most precise measurements at small angular scales to date—and in the Archeops balloon telescope.

Planck's Cosmic Microwave Background (CMB) Map

On 21 March 2013, the European-led research team behind the *Planck* cosmology probe released the mission's all-sky map (565 x 318 jpeg, 3600 x 1800 jpeg) of the *cosmic microwave background* [46,47]. See below.

[46] Clavin, W., & Harrington, J. D. (March 21, 2013). Planck Mission Brings Universe Into Sharp Focus. NASA.

[*According to the Big Bang theory*,] the map suggests the universe is slightly older than researchers expected.

[47] Staff (March 21, 2013). Mapping the Early Universe. *The New York Times*.

[*According to the Big Bang theory*,] subtle fluctuations in temperature were imprinted on the deep sky when the cosmos was about 370,000 years old. The imprint reflects ripples that arose as early, in the existence of the universe, as the first nonillionth (10−30) of a second. Apparently, these ripples gave rise to the present vast cosmic web of galaxy clusters and dark matter. Based on the 2013 data, the universe contains 4.9% *ordinary matter*, 26.8% *dark matter* and 68.3% *dark energy*. On February 5, 2015, new data was released by the Planck mission, according to which the *age of the universe* is 13.799 ± 0.021 billion years old and the *Hubble constant* was measured to be 67.74 ± 0.46 (km/s)/Mpc[48].

[48] Planck Collaboration (2016). Planck 2015 results. XIII. Cosmological parameters (See Table 4 on page 31 of pfd). *Astronomy & Astrophysics*, 594, 13, A13; arXiv:1502.01589; https://doi.org/10.1051/0004-6361/201525830.

Theoretical models

[*According to the Big Bang theory*,] the *cosmic microwave background* radiation and the *cosmological redshift-distance* relation are together regarded as the best available evidence for the *Big Bang event*.

Measurements of the CMB have made the inflationary *Big Bang model* the *Standard Cosmological Model* [50].

[50] Scott, D. (2005). The Standard Cosmological Model. *Canadian Journal of Physics*, 84, 6–7, 419–35; arXiv:astro-ph/0510731; https://doi.org/10.1139/P06-066.

The discovery of the *cosmic microwave background* (CMB) in the mid-1960s curtailed interest in alternatives, such as the *steady state theory*[51].

[51] Durham, F., & Purrington, R. D. (1983). *Frame of the universe: a history of physical cosmology*. Columbia University Press, pp. 193–209.

In the *Big Bang model* for the formation of the universe, inflationary cosmology predicts that after about 10^{-37} seconds[52] the nascent universe underwent exponential growth that smoothed out nearly all irregularities.

[52] Guth, A. H. (1998). *The Inflationary Universe: The Quest for a New Theory of Cosmic Origins*. Basic Books, p. 186.

[*According to the Big Bang theory*,] the remaining irregularities were caused by *quantum fluctuations* in the *inflation field* that caused the *inflation* event[53].

[53] Cirigliano, D., de Vega, H. J., & Sanchez, N. G. (2005). Clarifying inflation models: The precise inflationary potential from effective field theory and the WMAP data. *Phys. Rev.*, D (Submitted manuscript), 71, 10, 77–115; arXiv:astro-ph/0412634; https://doi.org/10.1103/PhysRevD.71.103518.

[*According to the Big Bang theory*,] long before the formation of *stars* and *planets*, the early universe was more compact, much hotter and, starting 10^{-6} seconds after the *Big Bang*, filled with a uniform glow from its white-hot fog of interacting *plasma* of *photons*, *electrons*, and *baryons*.

[*According to the Big Bang theory*,] as the universe expanded, adiabatic cooling caused the energy density of the *plasma* to decrease until it became favorable for *electrons* to combine with *protons*, forming *hydrogen atoms*. The *cosmic microwave background* (CMB) gives a snapshot of the hot early universe at the point in time when the temperature dropped enough to allow *electrons* and *protons* to form *hydrogen atoms*. This event made the universe nearly transparent to radiation because light was no longer being scattered off free *electrons*[62].

[62] Kaku, M. (2014). First Second of the Big Bang. How the Universe Works. Season 3. Episode 4. *Discovery Science*.

This corresponds to an ambient *energy* of about 0.26 eV, which is much less than the 13.6 eV ionization *energy* of hydrogen[63,64].

[63] Fixsen, D. J. (1995). *Formation of Structure in the Universe*. arXiv:astro-ph/9508159.
[64] Converted number: Conversion from K to eV.

[*According to the Big Bang theory,*] this epoch is known as the "*time of last scattering*" or the period of *recombination* or *decoupling*. This *recombination event* happened when the temperature was around 3,000 K or when the universe was approximately 379,000 years old[54].

[54] Abbott, B. (2007). Microwave (WMAP) All-Sky Survey. *Hayden Planetarium.*

As *photons* did not interact with these electrically *neutral* atoms, the former began to travel freely through space, resulting in the *decoupling* of *matter* and *radiation*[55].

[55] Gawiser, E., & Silk, J. (2000). The cosmic microwave background radiation. *Physics Reports*, 333–4, 245–267; arXiv:astro-ph/0002044; https://doi.org/10.1016/S0370-1573(00)00025-9.

[*According to the Big Bang theory,*] since *decoupling*, the *color temperature* of the background radiation has dropped by an average factor of 1,089 due to the *expansion of the universe*[40]. The *color temperature* of the ensemble of *decoupled photons* has continued to diminish ever since; now down to 2.7260 ± 0.0013 K[4], *it will continue to drop as the universe expands.* The *intensity* of the radiation corresponds to *black-body radiation* at 2.726 K *because red-shifted black-body radiation is just like black-body radiation at a lower temperature.*

[As mentioned above, this is a reversal of the previous argument that *the cosmological redshift was due to the expansion of space,* not due to the photons losing energy as they travel through *intergalactic space.*]

[*According to the Big Bang theory,*] the *color temperature* of this radiation stays inversely proportional to a parameter that describes the *relative expansion of the universe over time,* known as the *scale length.*]

The *color temperature* Tr of the *cosmic microwave background* (CMB) as a function of *redshift*, z, can be shown to be proportional to the *color temperature* of the CMB as observed in the present day (2.725 K or 0.2348 meV)[65]:

[65] Noterdaeme, P., Petitjean, P., Srianand, R., Ledoux, C., & López, S. (February, 2011). The evolution of the *cosmic microwave background* temperature. Measurements of TCMB at high *redshift* from carbon monoxide excitation. *Astronomy and Astrophysics*, 526, L7; arXiv:1012.3164; https://doi.org/10.1051/0004-6361/201016140.

$$Tr = 2.725 \text{ K} \times (1 + z)$$

[*According to the Big Bang model*,] the radiation from the sky we measure today comes from a spherical surface called the *surface of last scattering*. This represents the set of locations in space at which the *decoupling* event is estimated to have occurred[55,56] and at a point in time such that the *photons* from that distance have just reached observers.

[56] Smoot, G. F. (2006). *Cosmic Microwave Background Radiation Anisotropies: Their Discovery and Utilization*. Nobel Lecture. Nobel Foundation.
[57] NASA's "CMB Surface of Last Scatter".

Most of the *radiation energy* in the universe is in the *cosmic microwave background*[58], making up a fraction of roughly 6×10^{-5} of the total *density* of the universe.

[58] Hobson, M.P., Efstathiou, G., & Lasenby, A.N. (2006). *General Relativity: An Introduction for Physicists*. Cambridge University Press, pp. 388.
[59] Unsöld, A., & Bodo, B. (2002). *The New Cosmos, An Introduction to Astronomy and Astrophysics* (5th ed.). Springer-Verlag, p. 485.

[*According to the Big Bang theory*,] two of the greatest successes of the *Big Bang theory* are its prediction of the almost perfect *black body spectrum* and its detailed prediction of the *anisotropies* in the *cosmic microwave background.*

[*According to the Big Bang theory,*] t*he high degree of uniformity throughout the observable universe and its faint but measured anisotropy lend strong support for the Big Bang model in general and the ΛCDM ("Lambda Cold Dark Matter") model in particular.* [???] Moreover, the fluctuations are coherent on angular scales that are larger than the apparent cosmological horizon at recombination. Either such coherence is acausally fine-tuned, or cosmic inflation occurred[66,67].

> [66] Dodelson, S. (2003). Coherent Phase Argument for Inflation. *AIP Conference Proceedings.* 689, 184–96; arXiv:hep-ph/0309057. https://doi.org/10.1063/1.1627736.
> [67] Baumann, D. (2011). *The Physics of Inflation.* University of Cambridge.

The *cosmic microwave background* (CMB) spectrum has become the most precisely measured *black body spectrum* in nature[60].

> [60] White, M. (1999). Anisotropies in the CMB. *Proceedings of the Los Angeles Meeting, DPF 99.* UCLA; arXiv:astro-ph/9903232.

Predictions based on the Big Bang model

[*According to the Big Bang model,*] the *anisotropy,* or *directional dependency,* of the *cosmic microwave background* is divided into two types: *primary anisotropy,* due to effects that occur at the *surface of last scattering* and before; and *secondary anisotropy,* due to effects such as *interactions of the background radiation with intervening hot gas or gravitational potentials,* which occur between the *last scattering surface* and the observer.

The structure of the *cosmic microwave background anisotropies* is principally determined by two effects: *acoustic oscillations* and *diffusion damping* (also called *collisionless damping* or *Silk damping*).

[*According to the Big Bang model,*] the *acoustic oscillations* arise because of a conflict in the *photon–baryon plasma* in the *early universe*.

> [*Protons* and *neutrons* are *baryons*, a composite subatomic particle *that contains an odd number of valence quarks and antiquarks*, conventionally three. Because *quarks* have a *spin* ½, the difference in *quark* number results in being *fermions* because they have half-integer *spin*.]

The pressure of the *photons* tends to erase *anisotropies*, whereas the gravitational attraction of the *baryons*, moving at speeds much slower than light, makes them tend to collapse to form *overdensities*. These two effects compete to create *acoustic oscillations*, which give the *microwave background* its characteristic peak structure. The peaks correspond, roughly, to resonances in which the *photons* decouple when a particular mode is at its peak amplitude.

[*According to the Big Bang model,*] the peaks contain interesting physical signatures. The angular scale of the first peak determines the curvature of the universe (but not the topology of the universe).

The next peak—ratio of the odd peaks to the even peaks—determines the reduced baryon density[68].

[68] Hu, W. *Baryons and Inertia*.

[*According to the Big Bang model,*] the third peak can be used to get information about the *dark-matter* density.

[69] Hu, W. *Radiation Driving Force*.

The locations of the peaks give important information about the nature of the primordial *density perturbations*. There are two fundamental types of *density perturbations* called *adiabatic* and *isocurvature*. A general density perturbation is a mixture of both, and different theories that purport to explain the *primordial density perturbation spectrum* predict different mixtures.

In an *adiabatic density perturbation*, the fractional additional *number density* of each type of particle (*baryons, photons*, etc.) is the same. That is, if at one place there is a 1% higher *number density* of *baryons* than average, then at that place there is a 1% higher *number density* of *photons* (and a 1% higher *number density* in *neutrinos*) than average. *Cosmic inflation* predicts that the primordial perturbations are *adiabatic*.

In an *isocurvature density perturbation*, the sum (over different types of particle) of the fractional additional densities is zero. That is, a perturbation where at some spot there is 1% more *energy* in *baryons* than average, 1% more *energy* in *photons* than average, and 2% less *energy* in *neutrinos* than average, would be a pure *isocurvature perturbation*. Hypothetical *cosmic strings* would produce mostly *isocurvature primordial perturbations*.

The *cosmic microwave background* (CMB) spectrum can distinguish between these two because these two types of perturbations produce different peak locations. *Isocurvature density perturbations* produce a series of peaks whose angular scales (ℓ values of the peaks) are roughly in the ratio 1 : 3 : 5 : ..., while *adiabatic density perturbations* produce peaks whose locations are in the ratio 1 : 2 : 3 : ... .[70]

[70] Hu, W., & White, M. (1996). Acoustic Signatures in the Cosmic Microwave Background. *Astrophys. J.*, 471, 30–51; arXiv:astro-ph/9602019; https://doi.org/10.1086/177951.

[*According to the Big Bang model*,] observations are consistent with the *primordial density perturbations* being entirely *adiabatic*, providing key support for *inflation*, and ruling out many models of structure formation involving, for example, *cosmic strings*.

[*According to the Big Bang model*,] *collisionless damping* is caused by two effects, when the treatment of the primordial *plasma* as fluid begins to break down:

- the increasing mean free path of the *photons* as the primordial *plasma* becomes increasingly rarefied in an *expanding universe*,
- the finite depth of the *last scattering surface* (LSS), which causes the mean free path to increase rapidly during *decoupling*, even *while some Compton scattering is still occurring*.

[*According to the Big Bang model,*] these effects contribute about equally to the suppression of *anisotropies* at small scales and give rise to the characteristic exponential damping tail seen in the very small angular scale *anisotropies*.

The depth of the *last scattering surface* (LSS) refers to the fact that the *decoupling* of the *photons* and *baryons* does not happen instantaneously, but instead requires an appreciable fraction of the age of the universe up to that era. One method of quantifying how long this process took uses the *photon visibility function* (PVF). This function is defined so that, denoting the PVF by P(t), the probability that a *cosmic microwave background* (CMB) *photon* last *scattered* between time t and t + dt is given by P(t) dt. [NB].

[*According to the Big Bang model,*] the maximum of the PVF (the time when it is most likely that a given CMB *photon* last *scattered*) is known quite precisely. The first-year WMAP results put the time at which P(t) has a maximum as 372,000 years[71].

[71] WMAP Collaboration; Verde, L., *et al.* (2003). First-Year Wilkinson Microwave Anisotropy Probe (WMAP) Observations: Determination of Cosmological Parameters. *Astrophys. J. Supplement Series*, 148, 1, 175–94; arXiv:astro-ph/0302209; https://doi.org/10.1086/377226.

This is often taken as the "time" at which the *cosmic microwave background* (CMB) formed. However, to figure out how long it took the *photons* and *baryons* to decouple, we need a measure of the width of the *photon visibility function* (PVF). The WMAP team finds that the PVF is greater than half of its maximal value (the "full width at half maximum", or FWHM) over an interval of 115,000 years[71].

51

[*According to the Big Bang model,*] by this measure, *decoupling* took place over roughly 115,000 years, and thus when it was complete, the universe was roughly 487,000 years old.

Since the *cosmic microwave background* (CMB) came into existence, it has apparently been modified by several subsequent physical processes, which are collectively referred to as *late-time anisotropy*, or *secondary anisotropy*. When the CMB *photons* became free to travel unimpeded, ordinary matter in the universe was mostly in the form of *neutral hydrogen* and *helium atoms*. However, observations of galaxies today seem to indicate that most of the volume of the *intergalactic medium* (IGM) consists of ionized material [*plasma*] (*since there are few absorption lines due to hydrogen atoms*). This implies a period of *reionization* during which some of the material of the universe was broken into *hydrogen ions*.

The CMB *photons* are scattered by free charges such as *electrons* that are not bound in atoms. (NB). In an ionized universe, such *charged* particles have been liberated from neutral atoms *by ionizing (ultraviolet) radiation*. Today these free charges are at sufficiently low density in most of the volume of the universe that they do not measurably affect the CMB. However, if the IGM was ionized at very early times when the universe was still denser, then there are two main effects on the *cosmic microwave background* (CMB):
1. Small scale *anisotropies* are erased. (Just as when looking at an object through fog, details of the object appear fuzzy.)
2. The physics of how *photons* are scattered by free *electrons* (*Thomson scattering*) induces *polarization anisotropies* on large angular scales. This broad *angle polarization* is correlated with the *broad angle temperature perturbation*.

Both of these effects have been observed by the WMAP spacecraft, providing evidence that the universe was ionized at very early times, at a *redshift* around 10. [72]

[72] Hinshaw, G., *et al.* (September, 2013). Nine-year Wilkinson Microwave Anisotropy Probe (WMAP) Observations: Cosmological Parameter Results. *Astrophys. J. Supplement Series*, 208, 2, 19; arXiv:1212.5226; https://doi.org/10.1088/0067-0049/208/2/19.

The detailed provenance of this early ionizing radiation is still a matter of scientific debate. It may have included starlight from the very first population of *stars* (population III stars), *supernovae* when these first *stars* reached the end of their lives, or the ionizing radiation produced by the accretion disks of massive *black holes.*

The time following the emission of the *cosmic microwave background*—and before the observation of the first stars—is semi-humorously referred to by cosmologists as the *Dark Age*, and is a period which *is under intense study by astronomers.*

Two other effects which occurred between reionization and our observations of the *cosmic microwave background* and which appear to cause *anisotropies*, are the Sunyaev–Zeldovich effect, *where a cloud of high-energy electrons scatters the radiation*, transferring some of its energy to the CMB *photons*, and the Sachs–Wolfe effect, which causes *photons* from the *Cosmic Microwave Background* to be gravitationally redshifted or blueshifted due to changing gravitational fields.

Alternative theories

The *standard cosmology* that includes the *Big Bang* "enjoys considerable popularity among the practicing cosmologists"[73].

[73] Narlikar, J. V., & Padmanabhan, T. (September, 2001). Standard Cosmology and Alternatives: A Critical Appraisal. *Annual Review of Astronomy and Astrophysics*, 39, 1, 211–48; https://doi.org/10.1146/annurev.astro.39.1.211.

However, *there are challenges to the standard Big Bang framework for explaining cosmic microwave background* (CMB) *data*. In particular *standard cosmology* requires fine-tuning of some free parameters, with different values supported by different experimental data[73]. As an example of the fine-tuning issue, *standard cosmology cannot predict the present temperature of the relic radiation, T$_0$.*[73] *This value of T$_0$ is one of the best results of experimental cosmology and the steady state model can predict it* [61].

[61] Assis, A. K. T., Paulo, S., & Neves, M. C. D. (July, 1995). History of the 2.7 K Temperature Prior to Penzias and Wilson". *Apeiron*, 2, 3, 79–87.

However, alternative models have their own set of problems and *they have only made post-facto explanations of existing observations*[73] [??? See below.] Nevertheless, these alternatives have played an important historic role in providing ideas for and challenges to the standard explanation[12].

Data analysis challenges

Raw *cosmic microwave background radiation* (CMBR) data, even from space vehicles such as WMAP or *Planck*, contain foreground effects that completely obscure the fine-scale structure of the *cosmic microwave background*. The fine-scale structure is superimposed on the raw CMBR data but is too small to be seen at the scale of the raw data. The most prominent of the foreground effects is the *dipole anisotropy* caused by the *Sun's motion relative to the CMBR background*. The *dipole anisotropy* and others due to *Earth's annual motion relative to the Sun* and numerous microwave sources in the galactic plane and elsewhere must be subtracted out to reveal the extremely tiny variations characterizing the fine-scale structure of the CMBR *background*. The detailed analysis of CMBR data to produce maps, an *angular power spectrum*, and ultimately cosmological parameters is a complicated, computationally difficult problem.

In practice it is hard to take the effects of noise and foreground sources into account. In particular, these foregrounds are dominated by galactic emissions such as *Bremsstrahlung, synchrotron,* and *dust* that *emit in the microwave band*; in practice, the galaxy has to be removed, resulting in a *cosmic microwave background* (CMB) map that is not a full-sky map. In addition, point sources like *galaxies* and *clusters* represent another source of foreground which must be removed so as not to distort the short scale structure of the CMB power spectrum.

[*Synchrotron radiation* (also known as *magnetobremsstrahlung*) is the *electromagnetic radiation* emitted when *relativistic charged particles* are subject to an acceleration perpendicular to their velocity (a ⊥ v). It is produced artificially in some types of particle accelerators or naturally by fast *electrons* moving through *magnetic fields*. The radiation produced in this way has a characteristic *polarization*, and the *frequencies* generated can range over a large portion of the electromagnetic spectrum.

Synchrotron radiation is similar to *bremsstrahlung radiation*, which is emitted by a *charged particle* when the acceleration is parallel to the direction of motion. The general term for radiation emitted by particles in a *magnetic field* is *gyromagnetic radiation*, for which *synchrotron radiation* is the *ultra-relativistic* special case. Radiation emitted by charged particles moving *non-relativistically* in a *magnetic field* is called *cyclotron emission*. For particles in the *mildly relativistic range* (≈ 85% of the speed of light), the *emission* is termed *gyro-synchrotron radiation*.

In *astrophysics*, *synchrotron emission* occurs, for instance, due to *ultra-relativistic motion* of a *charged particle* around a *black hole*. When the source follows a circular geodesic around the *black hole*, the *synchrotron radiation* occurs for orbits close to the photosphere where the motion is in the *ultra-relativistic* regime.]

Constraints on many cosmological parameters can be obtained from their effects on the power spectrum, and results are often calculated using Markov chain Monte Carlo sampling techniques.

With the increasingly precise data provided by WMAP, there have been a number of claims that the CMB *exhibits anomalies*, such as very large-scale *anisotropies*, anomalous alignments, and non-Gaussian distributions[92,93,94].

[92] Rossmanith, G., Räth, C., Banday, A. J., & Morfill, G. (2009). Non-Gaussian Signatures in the five-year WMAP data as identified with isotropic scaling indices. *MNRAS*, 399, 4, 1921–33; arXiv:0905.2854; https://doi.org/10.1111/j.1365-2966.2009.15421.x.

[93] Bernui, A., Mota, B., Rebouças, M. J., & Tavakol, R. (2007). Mapping the large-scale anisotropy in the WMAP data. *Astronomy and Astrophysics*, 464, 2, 479–85; arXiv:astro-ph/0511666; https://doi.org/10.1051/0004-6361:20065585.

[94] Jaffe, T.R., Banday, A. J., Eriksen, H. K., Górski, K. M., & Hansen, F. K. (2005). Evidence of vorticity and shear at large angular scales in the WMAP data: a violation of cosmological isotropy? *Astrophys. J.*, 629, 1, L1–L4; arXiv:astro-ph/0503213; https://doi.org/10.1086/444454.

The most longstanding of these is the *low-ℓ multipole controversy*. Even in the COBE map, it was observed that *the quadrupole (ℓ = 2, spherical harmonic) has a low amplitude compared to the predictions of the Big Bang*. In particular, the quadrupole and octupole (ℓ = 3) modes appear to have *an unexplained alignment with each other and with both the ecliptic plane and equinoxes*[95,96,97].

[95] de Oliveira-Costa, A., Tegmark, M., Zaldarriaga, M., & Hamilton, A. (2004). The significance of the largest scale CMB fluctuations in WMAP. *Phys. Rev.*, D (Submitted manuscript), 69, 6, 063516; arXiv:astro-ph/0307282;https://doi.org/10.1103/PhysRevD.69.063516.

[96] Schwarz, D. J., Starkman, G. D., *et al.* (2004). Is the low-ℓ microwave background cosmic? *Phys. Rev. Lett.*, (Submitted manuscript), 93, 22, 221301; arXiv:astro-ph/0403353; https://doi.org/10.1103/PhysRevLett.93.221301.

[97] Bielewicz, P., Gorski, K. M., & Banday, A. J. (2004). Low-order multipole maps of CMB anisotropy derived from WMAP. *MNRAS*,

355, 4, 1283–302; arXiv:astro-ph/0405007; https://doi.org/10.1111/
j.1365-2966.2004.08405.x.

A number of groups have suggested that this could be the signature of *new physics* at the greatest observable scales; other groups suspect systematic errors in the data[98,99,100].

[98] Liu, H., & Li, T-P. (2009). *Improved CMB Map from WMAP Data*; arXiv:0907.2731v3 [astro-ph].
[99] Sawangwit, U., & Shanks, T. (2010). *Lambda-CDM and the WMAP Power Spectrum Beam Profile Sensitivity*; arXiv:1006.1270v1 [astro-ph].
[100] Liu, H., *et al*. (2010). Diagnosing Timing Error in WMAP Data. *MNRAS, Letters*, 413, 1, L96–L100; arXiv:1009.2701v1; https://doi.org/10.1111/j.1745-3933.2011.01041.x.

Ultimately, due to the foregrounds and the cosmic variance problem, the greatest modes will never be as well measured as the small angular scale modes. The analyses were performed on two maps that have had the foregrounds removed as far as possible: the "internal linear combination" map of the WMAP collaboration and a similar map prepared by Max Tegmark and others[101,40,102].

[101] Hinshaw, G., *et al*. (WMAP collaboration) (2007). Three-year Wilkinson Microwave Anisotropy Probe (WMAP) observations: temperature analysis. *Astrophys. J. Supplement Series*, 170, 2, 288–334; arXiv:astro-ph/0603451; https://doi.org/10.1086/513698.
[102] Tegmark, M., de Oliveira-Costa, A., & Hamilton, A. (2003). A high-resolution foreground cleaned CMB map from WMAP. *Phys. Rev.*, D, 68, 12, 123523; arXiv:astro-ph/0302496; https://doi.org/10.1103/PhysRevD.68.123523. This paper states, "Not surprisingly, the two most contaminated multipoles are [the quadrupole and octupole], which most closely trace the galactic plane morphology."

Later analyses have pointed out that these are the modes most susceptible to foreground contamination from *synchrotron*, *dust*, and

Bremsstrahlung emission, and from *experimental uncertainty* in the *monopole* and *dipole*.

A full Bayesian analysis of the WMAP power spectrum demonstrates that the quadrupole prediction of *Lambda-CDM cosmology* is consistent with the data at the 10% level and that the observed octupole is not remarkable[103].

[103] O'Dwyer, I., *et al.* (2004). Bayesian Power Spectrum Analysis of the First-Year Wilkinson Microwave Anisotropy Probe Data. *Astrophys. J. Letters*, 617, 2, L99–L102; arXiv:astro-ph/0407027; https://doi.org/ 10.1086/427386.

Carefully accounting for the procedure used to remove the foregrounds from the full sky map further reduces the significance of the alignment by ~5% [104,105,106,107].

[104] Slosar, A., & Seljak, U. (2004). Assessing the effects of foregrounds and sky removal in WMAP. *Phys. Rev.*, D, (Submitted manuscript), 70, 8, 083002; arXiv:astro-ph/0404567; https://doi.org/ 10.1103/PhysRevD.70.083002.
[105] Bielewicz, P., Eriksen, H. K., Banday, A. J., Górski, K. M., & Lilje, P. B. (2005). Multipole vector anomalies in the first-year WMAP data: a cut-sky analysis. *Astrophys. J.*, 635, 2, 750–60; arXiv:astro-ph/0507186; https://doi.org/10.1086/497263.
[106] Copi, C. J., Huterer, D., Schwarz, D. J., & Starkman, G. D. (2006). On the large-angle anomalies of the microwave sky. *MNRAS*, 367, 1, 79–102; https://doi.org/10.1111/j.1365-2966.2005.09980.x
[107] de Oliveira-Costa, A., & Tegmark, M. (2006). CMB multipole measurements in the presence of foregrounds. *Phys. Rev.*, D, (Submitted manuscript), 74, 2, 023005; arXiv:astro-ph/0603369; https://doi.org/10.1103/PhysRevD.74.023005.

Recent observations with the *Planck* telescope, which is very much more sensitive than WMAP and has a larger angular resolution, record *the same anomaly, and so instrumental error (but not foreground contamination) appears to be ruled out.*

Coincidence is a possible explanation. Chief scientist from WMAP, Charles L. Bennett suggested coincidence and human psychology were involved, "I do think there is a bit of a psychological effect; people want to find unusual things".

Measurements of the *density of quasars* based on Wide-field Infrared Survey Explorer data *finds a dipole significantly different from the one extracted from the CMB anisotropy*[110].

[110] Secrest, N. J., von Hausegger, S., Rameez, M., Mohayaee, R., Sarkar, S., & Colin, J. (2021). A Test of the Cosmological Principle with Quasars. *Astrophys. J. Lett.*, 908, 2, L51; arXiv:2009.14826; https://doi.org/10.3847/2041-8213/abdd40.

This difference is in conflict with the cosmological principle[111].

[111] Perivolaropoulos, L., & Skara, F. (December, 2022). Challenges for ΛCDM: An update. *New Astronomy Reviews*, 95, 101659; arXiv:2105.05208; https://doi.org/10.1016/j.newar.2022.101659.

[The *cosmological principle* states that the universe is homogeneous and isotropic on a large scale. This means that the distribution of *matter* and *energy* is uniform, and the universe looks the same in all directions. It asserts that there are no preferred locations or directions in space.]

[*According to the Big Bang model,*] *assuming the universe keeps expanding* and it does not suffer a *Big Crunch*, a *Big Rip*, or another similar fate, the *cosmic microwave background* will continue *red shifting* until it will no longer be detectable[112], and will be superseded first by *the one produced by starlight*, and perhaps, later by the background radiation fields of processes that may take place in the far future of the universe such as *proton decay, evaporation of black holes,* and *positronium decay*[113].

[112] Krauss, L. M., & Scherrer, R. J. (2007). The return of a static universe and the end of cosmology. *General Relativity and*

Gravitation, 39, 10, 1545–50; arXiv:0704.0221; https://doi.org/10.1007/s10714-007-0472-9.

[113] Adams, F. C., & Laughlin, G. (1997). A dying universe: The long-term fate and evolution of astrophysical objects. *Rev. Mod. Phys.*, 69, 2, 337–72; arXiv:astro-ph/9701131; https://doi.org/10.1103/RevModPhys.69.337.

Hannes Olof Gösta Alfvén (May 30, 1908–April 2, 1995).

Alfvén was a Swedish electrical engineer, plasma physicist and winner of the 1970 Nobel Prize in Physics for his work on *magnetohydrodynamics* (MHD). He described the class of MHD waves now known as Alfvén waves. He was originally trained as an electrical power engineer and later moved to research and teaching in the fields of *plasma physics* and *electrical engineering*.

[*Plasma* is one of four fundamental states of *matter* (the other three being *solid*, *liquid*, and *gas*) characterized by the *presence of a significant portion of charged particles in any combination of ions or electrons*. It is the most abundant form of ordinary *matter* in the universe, mostly in *stars* (including the Sun), but also dominating the rarefied *intracluster medium* and *intergalactic medium*. *Plasma* can be artificially generated, for example, by heating a neutral gas or subjecting it to a strong electromagnetic field. The presence of *charged particles* makes *plasma electrically conductive*, with the dynamics of individual particles and macroscopic *plasma* motion governed by collective *electromagnetic fields* and very sensitive to externally applied fields.]

Alfvén made many contributions to *plasma physics*, including *theories describing the behavior of aurorae*, the *Van Allen radiation belts*, the *effect of magnetic storms on the Earth's magnetic field*, the *terrestrial magnetosphere*, and the *dynamics of plasmas in the Milky Way galaxy*.

Alfvén received his PhD from the University of Uppsala in 1934. His thesis was titled "*Investigations of High-frequency Electromagnetic Waves*." In 1934, Alfvén taught physics at both the University of Uppsala and the Nobel Institute for Physics (later renamed the Manne Siegbahn Institute of Physics) in Stockholm, Sweden. In 1940, he became professor of electromagnetic theory and electrical measurements at the Royal Institute of Technology in Stockholm. In

1945, he acquired the non-appointive position of Chair of Electronics. His title was changed to *Chair of Plasma Physics* in 1963. From 1954 to 1955, Alfvén was a Fulbright Scholar at the University of Maryland, College Park.

In 1937, Alfvén argued that if *plasma* pervaded the universe, it could then carry electric currents capable of generating a galactic magnetic field. [Alfvén, H. (1937). Cosmic Radiation as an Intra-galactic Phenomenon. *Arkiv För Matematik, Astronomi Och Fysik*, 25B, 29, Uppsala.]

Alfvén believed the problem with the Big Bang was that astrophysicists tried to extrapolate the origin of the universe from mathematical theories developed on the blackboard, rather than starting from known observable phenomena. Alfvén and colleagues proposed the Alfvén–Klein model as an alternative cosmological theory to both the Big Bang and steady state theory cosmologies.

In 1939, Alfvén proposed the theory of magnetic storms and auroras and the theory of *plasma* dynamics in the Earth's magnetosphere. This paper was rejected by the U.S. journal *Terrestrial Magnetism and Atmospheric Electricity*. [Alfvén, H. (1937). A theory of magnetic storms and of the aurorae. *Kungl. Svenska Vetenskapsakademiens Handlingar*, Third Series, 18, 3 & 9, Stockholm.]

In 1942 he published a letter to *Nature* "*Existence of electromagnetic-hydrodynamic waves.*" [Alfvén, H. (1942). Existence of electromagnetic-hydrodynamic waves. *Nature*, 150, 405-6.; https://doi.org/10.1038/ 150405d0.] The letter, which was only half a page long, described a new type of low-frequency oscillation of a magnetized *plasma*. The paper was long disregarded or openly rejected, as the new waves that it described could not be demonstrated experimentally at that time. But as experimental techniques improved in the laboratory and experiments in space became possible, it became evident that what are now called Alfvén waves were of fundamental importance in *plasma* physics in general and *space plasma physics* in particular. In fact, they constitute a cornerstone of a new field of physics: *magneto hydrodynamics*. Alfvén waves are found in just about every space *plasma*, such as the Sun, the solar corona, and the solar wind, as well as in magnetospheres of the Earth and other magnetized planets.

In 1954 he published *"On the origin of the solar system"*. Oxford, Clarendon Press. (See below.) In 1967, after leaving Sweden and spending time in the Soviet Union, he moved to the United States. Alfvén worked in the departments of electrical engineering at both the University of California, San Diego and the University of Southern California.

In December, 1970, he was awarded the Nobel Prize in Physics 1970 "for fundamental work and discoveries in *magnetohydro-dynamics* with fruitful applications in different parts of *plasma physics*".

Many of his theories about the solar system were verified as late as the 1980s through external measurements of cometary and planetary magnetospheres. However, Alfvén himself noted that astrophysical textbooks poorly represented known *plasma* phenomena. Many physicists regarded Alfvén as espousing unorthodox opinions. Alfvén recalled: "When I describe [*plasma* phenomena] according to this formalism most referees do not understand what I say and turn down my papers. With the referee system which rules US science today, this means that my papers are rarely accepted by the leading US journals.

In 1991, Alfvén retired as professor of electrical engineering at the University of California, San Diego and professor of plasma physics at the Royal Institute of Technology in Stockholm. The asteroid 1778 Alfvén is named in his honor.

Alfvén was married for 67 years to his wife Kerstin (1910–1992). They raised five children, one boy and four girls. Their son became a physician, while one daughter became a writer and another a lawyer in Sweden. He spent his later adult life alternating between California and Sweden. Alfvén died on April 2, 1995, at Djursholm, aged 86.

Alfvén, H. (1937). Cosmic Radiation as an Intra-galactic Phenomenon.

Arkiv För Matematik, Astronomi Och Fysik, 25B, 29, Uppsala.

Alfvén argued that if *plasma* pervaded the universe, it could then carry *electric currents* capable of generating a *galactic magnetic field*.

Alfvén, H. (1937) A theory of magnetic storms and of the aurorae.

Kungl. Svenska Vetenskapsakademiens Handlingar, Third Series, 18, 3 & 9, Stockholm.

Alfvén attributes *magnetic storms* and *auroras* to a stream of *ions* and *electrons* emitted from the *Sun*, the *electrons* with energies about 10^8 volts and the ions with smaller energies. The theory differs from Chapman and Ferraro's theory of magnetic storms by taking into account the *magnetic field* of the *Sun* and supposing that the motion of any individual particle is hardly affected by the *magnetic field* due to the motion of other particles. The outward motion of the stream from Sun to Earth is explained as due, not to the original outward velocities of the particles, but to *electromagnetic effects*.

Alfvén, H. (1942). Existence of electromagnetic-hydrodynamic waves.

Nature, 150, 405-6.; https://doi.org/10.1038/ 150405d0.

If a *conducting* liquid is placed in a constant *magnetic field*, every motion of the liquid gives rise to an *electromagnetic field* which produces *electric currents*. Owing to the *magnetic field*, these *currents* give mechanical *forces* which change the state of motion of the liquid. Thus, a kind of combined *electromagnetic-hydro-dynamic wave* is produced which, so far as I know, has as yet attracted no attention.

Alfvén, H. (1954). *On the origin of the solar system"*.

Oxford, Clarendon Press.

Preface

The theory of the origin of the *solar system* which is presented here consists essentially of two parts which to a certain extent are independent. The first part, developed in Chapters I-IV, is a general theory of the *formation* and gives as a result the *mass distribution in the solar system*; the second part, contained in Chapters V-IX, deals with some details of the *structure*. Chapters X and XI discuss some general aspects and are more directly related to the first part. Finally, Chapter XII gives a summary of the results.

It is an interesting question how much of the detail of the *solar system* a cosmogonic theory should attempt to explain. It seems to be quite a reasonable view that we could never hope, and should not even try, to make a theory to explain more than the most general features of the system. So long a time has elapsed since the formation, and so many changes in the constitution may have occurred, that it would not be advisable to consider the details. During the work in this field, I have become convinced of the opposite: *a theory of the origin could, and should, be carried out even to the details*. This is the reason for including Chapters V-IX in this treatise. Many readers will probably disagree on this point. They are advised to pass directly from Chapter IV to Chapter X, or at least to avoid Chapters VI and VII.

Innumerable discussions with Dr. Herlofson have been most valuable to me. I also wish to thank Dr. Lundquist and Mr. AstrOm for much advice, especially concerning Chapters V and VIII. Mr. Ekstrand has kindly helped me with numerical calculations. …

H. A.
STOCKHOLM
ROYAL INSTITUTE OF TECHNOLOGY
31 October 1952

Erickson, W. C. (November, 1957). A Mechanism of Non-Thermal Radio-Noise Origin*.

Astrophys. J., 126, 480; https://adsabs.harvard.edu/pdf/ 1957ApJ...126..480E.

* Based on a portion of a thesis submitted to the physics department of the University of Minnesota in partial fulfilment of the requirements for the degree of Doctor of Philosophy.

The author also wishes to acknowledge the financial assistance given to him by the National Science Foundation; he held a National Science Foundation Pre-doctoràl Fellowship during the preliminary stages of this work.

Department of Physics, University of Minnesota, and Carnegie Institution of Washington. Department of Terrestrial Magnetism, Washington 15, D.C.

Received September 19, 1956.
Revised May 10, 1957.

―――――――――――――

Abstract

A mechanism of *non-thermal radio-noise* origin is proposed.

[*Cosmic noise*, also known as *galactic radio noise*, is a physical phenomenon derived from outside of the Earth's atmosphere. Its characteristics are comparable to those of *thermal noise*. *Cosmic noise* occurs at *frequencies* above about 15 MHz when highly directional antennas are pointed toward the Sun or other regions of the sky, such as the center of the Milky Way Galaxy. Celestial objects like *quasars*, which are super dense objects far from Earth, emit *electromagnetic waves* in their full spectrum, including *radio waves*. *Cosmic microwave background radiation* (CMBR) *from outer space is also a form of cosmic noise. CMBR is thought to be a relic of the Big Bang, and pervades the space almost homogeneously over the entire celestial sphere.* The bandwidth of the *cosmic microwave background radiation* (CMBR) is wide, though the peak is in the *microwave range.*]

The action of this mechanism may be summarized in the following manner. Suppose that clouds of *interstellar grains* exist in the *radio-source* regions. If a *high-velocity gas cloud* collides with a cloud of *grains*, the *grains* will be bombarded by moderately fast *atoms* and/or *ions*. These collisions will transfer *angular momentum* to the grains, and, in fact, the *angular velocity* of each grain will execute a dynamical "walk." *It is shown that rotational frequencies comparable with radio frequencies may be attained. If some of the grains possess electric or magnetic dipole moments due to polar or ferromagnetic substances or statistical fluctuations in the distribution of charge on the grains, they will radiate classically at radio frequencies.* Rather improbably high grain densities are required in order to account for the *total radio-frequency radiation of high-emissivity sources.* However, *the high-frequency portion of this radiation could be generated with moderate grain densities.*

I. INTRODUCTION

It is well known that *discrete radio-noise sources appear to be composed of clouds of rarified gases possessing enormous velocity dispersions.* Baade & Minkowski, (1954)[1,2] have shown that the clouds possess random velocities of 300-3000 km/sec with respect to one another.

[1] Baade, W., & Minkowski, R. (January, 1954). Identification of the Radio Sources in Cassiopeia, Cygnus A, and Puppis A. *Astrophys. J.*, 119, 206; https://doi.org/10.1086/145812;

[2] Baade, W., & Minkowski, R. (1954). On the Identification of Radio Sources. *Ibid.*, 119, 215; https://doi.org/10.1086/145813.

Minkowski & Aller, (1954)[3] have examined the optical spectrum of the *Cassiopeia A* source.

[3] Minkowski, R, & Aller, L. H. (1954). The Spectrum of the Radio Source in Cassiopeia. *Astrophys. J.*, 119, 232; https://adsabs.harvard.edu/full/1954ApJ...119..232M.

They find no reason to assume an abnormal chemical composition of the gas. Therefore, it can be assumed to be principally *hydrogen*. Their estimate of the *electron* density is 10^4-10^5 cm^{-3}.

If *interstellar grains* exist in *radio-source regions, collisions* with the high-velocity gas will excite them to states of rapid rotation. In fact, it will be shown that they will rotate at *radio frequencies*. Thus, *if an appreciable number of the grains possess electric or magnetic moments, they will radiate classically at radio frequencies*. It can be shown that, for the range of *angular velocities* of the grains and the *translational velocities* of the gas under consideration, equipartition of energy between the rotational degrees of freedom of the grains and the translational degrees of freedom of the gas cannot always be assumed. Therefore, the interaction between the gas and the grains must be examined in greater detail. *It is found that the interaction is insensitive to the degree of ionization of the hydrogen gas. The electrons of the gas, whether bound or unbound, may be neglected, and only the interaction between the protons and the grains must be considered.*

For calculational purposes, it will be assumed that the *grains* are spherical. The assumption of non-spherical *grains* requires a far more complex calculation than would appear to be justified at the present time. The *thermal* and *turbulent* velocities of the *protons* will be neglected, as compared to their *translational velocities*.

In *collisions with the grains*, the *protons* can transfer no *angular momentum in the direction parallel to their velocity*; thus, the *angular momentum is restricted to the plane perpendicular to the proton velocities*, and the *angular velocities* of the *grains* each execute a two-dimensional dynamical "walk" as the *grains* are struck successively by the *protons*. Because of the radiation reaction and for other reasons, various torques will act on each *grain*. Thus, various drifts are superimposed on the dynamical walk of the *angular velocity*.

II. INTERACTION OE THE GRAINS WITH THE GAS

a) General Equations

Consider a *cloud of interstellar grains*. Before this *cloud* is struck by a *high-velocity gas cloud*, the *grains* may be considered to be essentially at rest. During the collision the *grains* are bombarded by *protons* with velocities on the order of 10^8 cm/sec (5-kev energies). Owing to this bombardment, the *grains* acquire both *angular* and *linear momentum*. If the gas cloud is large enough, the *grains* will eventually acquire the same *velocity* as the *protons*; thus, their *translational kinetic energy* approaches $\frac{1}{2} MV^2$, where M is the *mass* of a typical *grain* and V is the gas, or *proton*, *velocity*. It will be shown that during the *acceleration process* the *rotational kinetic energy* of the *grains* approaches the order of magnitude of $\frac{1}{2} mV^2$, where m is the *proton mass*. Eventually, as the *grain* acquires the same *velocity* as the *gas*, the bombardment ceases, and various dissipative processes cause the *rotational kinetic energy* of the *grains* to decrease to approximately kT, where T is the *temperature* of the *gas*. *It is during the acceleration process, when $E_{rot} \sim \frac{1}{2} mV^2$, that rotational frequencies comparable with radio frequencies are attained.* Suppose that the *average momentum* transferred to a *grain* due to a *proton collision* is p and that the *average impact parameter* for a collision is *a*. Then the change in the *angular momentum* of the *grain* is p*a*. If Ω denotes the *change in the angular velocity* and I is the *moment of inertia* of a typical *grain*, p*a* = $I\Omega$. Since this is essentially a *dynamical walk process*, after N such collisions the mean-square *angular velocity* of the *grain* is given by

$$\omega^2 \simeq N\Omega^2 = (pa/I)^2; \qquad (1)$$

thus, the *rotational kinetic energy* of the *grain* is

$$E_{rot} = \frac{1}{2} I\omega^2 = Np^2a^2/2I. \qquad (2)$$

On the other hand, after N such collisions the *linear momentum* of the *grain* is Np; thus, the *translational kinetic energy* of the *grain* is

$$E_{tran} = Np^2/2M. \qquad (3)$$

For *spherical grains*, a is very nearly equal to the *radius of gyration* of the *grain*; therefore,

$$E_{rot}/E_{tran} \simeq 1/N \ (Ma^2/I) \simeq 1/N. \qquad (4)$$

It will be shown in Section IIb that *p is quite insensitive to the velocity of the proton relative to the grain* in the range of 10^7-10^9 cm/sec; thus, p is fairly constant over most of the time during which the *grain* is being accelerated, and a constant β may be defined by the relation $p = \beta mV$ (V is the *proton velocity* in the frame of reference in which the grain was initially at rest). As $E_{tran} \rightarrow \frac{1}{2} MV^2$, $N \rightarrow MV/p = M/m\beta$, and thus, from equation (4), $E_{rot} \rightarrow m\beta/M(\frac{1}{2} MV^2) = \beta \ (\frac{1}{2} mV^2)$. It will be shown that β, the fraction of its *momentum* which a *proton* loses in passing through a typical *grain*, is less than unity but not less than 1/10. Therefore, $\beta < 1$, and during the *acceleration* process E_{rot} approaches the order of magnitude of $\frac{1}{2} mV^2$. If the *gas clouds* are small enough that the *grains* do not get carried along with them, a typical *grain* would be successively bombarded by *protons* of clouds traveling in various directions; under such conditions, $E_{rot} \sim E_{tran} \sim \frac{1}{2} mV^2$, i.e., *equipartion* will probably ensue. ...

[In classical statistical mechanics, the *equipartition* theorem relates the temperature of a system to its average *energies*. The *equipartition theorem* is also known as the *law of equipartition, equipartition of energy*, or simply *equipartition*. The original idea of equipartition was that, *in thermal equilibrium, energy is shared equally among all of its various forms*; for example, *the average kinetic energy per degree of freedom in translational motion of a molecule should equal that in rotational motion*.]

V. CONCLUSIONS

It would appear possible that a mechanism similar to the one discussed here could be responsible for the production of at least part of the observed non-thermal radio noise. If this mechanism is to be responsible for *all the non-thermal radio-noise production,* the requirements on the *mass density* of *grains* and the *dipole moments* of *grains* become quite stringent. It should be noted that the *radiation densities* in the source regions are high, so that ... the *coefficient for induced emission and absorption* is far larger than the *coefficient for spontaneous emission of radiation.* Therefore, *if the population of the upper rotational-energy levels of the grains were to exceed that of the lower levels,* MASER-like action [*microwave amplification by stimulated emission of radiation*] would ensue, and *large amounts of radiation would be produced.* The *mass-density* and *dipole-moment* requirements would then become trivial.

[A *maser* is a device that produces coherent *electromagnetic waves* in the *microwave* region of the *electromagnetic spectrum* (*microwaves*), through amplification by stimulated emission.

The *maser* is based on the *principle of stimulated emission* proposed by Albert Einstein in 1917. When atoms have been induced into an *excited energy state*, they can amplify radiation at a frequency particular to the element or molecule used as the masing medium. By putting such an amplifying medium in a resonant cavity, feedback is created that can produce coherent radiation.

Nikolay Basov, Alexander Prokhorov and Joseph Weber introduced the concept of the maser in 1952, and Charles H. Townes, James P. Gordon, and Herbert J. Zeiger built the first maser at Columbia University in 1953. Townes, Basov and Prokhorov won the 1964 Nobel Prize in Physics for theoretical work leading to the maser.

71

Maser-like stimulated emission has also been observed in nature from *interstellar space*, and it is frequently called "super-radiant emission" to distinguish it from laboratory masers. Such emission is observed from molecules such as water (H_2O), hydroxyl radicals (•OH), methanol (CH_3OH), formaldehyde (HCHO), silicon monoxide (SiO), and carbodiimide (HNCNH). Water molecules in star-forming regions can undergo a population inversion and emit radiation at about 22.0 GHz, creating the brightest spectral line in the radio universe. Some water masers also emit radiation from a rotational transition at a frequency of 96 GHz. Extremely powerful masers, associated with *active galactic nuclei*, are known as *megamasers* and are up to a million times more powerful than *stellar* masers.]

Dicke, R. H., Peebles, P. J. E., Roll, P. G., & Wilkinson, D. T. (July, 1965). Cosmic Black-body Radiation.

Astrophys. J., 142, 414; https://articles.adsabs.harvard.edu/pdf/ 1965ApJ...142..414D.

* This research was supported in part by the National Science Foundation and the Office of Naval Research of the U.S. Navy.

Submitted May 7, 1965.

Palmer Physical Laboratory Princeton, New Jersey.

One of the basic problems of cosmology is the *singularity characteristic* of the familiar cosmological solutions of *Einstein's field equations.* Also puzzling, is the presence of *matter* in excess over *antimatter* in the universe, for *baryons* and *leptons* are thought to be conserved. *Thus, in the framework of conventional theory we cannot understand the origin of matter or of the universe.* We can distinguish three main attempts to deal with these problems.

1. The assumption of continuous creation (Bondi & Gold, 1948[1]; Hoyle, 1948[2]), *which avoids the singularity by postulating a universe expanding for all time and a continuous but slow creation of new matter in the universe.*

[1] Bondi, H, & Gold, T. (June, 1948). The Steady-State Theory of the Expanding Universe. MNRAS, 108, 3, 252–70; https://doi.org/ 10.1093/mnras/108.3.252.
[2] Hoyle, F. (October, 1948). A New Model for the Expanding Universe. *MNRAS*, 108, 5, 372-82; https://doi.org/10.1093/mnras/ 108.5.372

2. The assumption (Wheeler, 1964[3]) that *the creation of new matter is intimately related to the existence of the singularity,* and that the

resolution of both paradoxes may be found in a proper quantum mechanical treatment of *Einstein's field equations*.

[3] Wheeler, J. A. (1964), in *Relativity, Groups and Topology*, ed C. DeWitt and B. DeWitt, Gordon & Breach, New York.

3. The assumption that *the singularity results from a mathematical over-idealization, the requirement of strict isotropy or uniformity*, and that it would not occur in the real world (Wheeler, 1958[4]; Lifshitz & Khalatnikov, 1963[5]).

[4] Wheeler, J. A. (1958). *La Structure et l'evolution de l'universe*. 11th Solvay Conf., Éditions Stoops, Brussels, p. 112.
[5] Liftshitz, E M., & Khalatnikov, I. M. (April, 1963). Investigations in relativistic cosmology. *Adv. in Phys.*, 12, 185-249; https://doi.org/10.1080/00018736300101283.

If this third premise is accepted tentatively as a working hypothesis, it carries with it a possible resolution of the second paradox, for the *matter* we see about us now may represent the same *baryon* content of the previous expansion of a closed universe, oscillating for all time. This relieves us of the necessity of understanding the origin of *matter* at any finite time in the past. In this picture it is essential to suppose that at the time of maximum collapse the temperature of the universe would exceed 10^{10} °K, in order that the ashes of the previous cycle would have been reprocessed back to the hydrogen required for the *stars* in the next cycle.

Even without this hypothesis it is of interest to inquire about the temperature of the universe in these earlier times. From this broader viewpoint we need not limit the discussion to closed oscillating models. Even if the universe had a singular origin, it might have been extremely hot in the early stages.

Could the universe have been filled with *black-body radiation* from this possible high-temperature state? If so, it is important to notice that as the universe *expands* the *cosmological redshift* would serve to adiabatically cool the radiation, while preserving the thermal character. The radiation temperature would vary inversely as the *expansion* parameter (radius) of the universe.

The presence of thermal radiation remaining from the fireball is to be expected if we can trace the *expansion* of the universe back to a time when the temperature was of the order of 10^{10} °K ($\sim m_e c^2$). In this state, we would expect to find that the *electron* abundance had increased very substantially, due to thermal *electron-pair* production, to a density characteristic of the temperature only. One readily verifies that, whatever the previous history of the universe, the *photon* absorption length would have been short with this high *electron density*, and the radiation content of the universe would have promptly adjusted to a thermal equilibrium distribution due to *pair-creation* and *annihilation* processes. This adjustment requires a time interval short compared with the characteristic *expansion time* of the universe, whether the cosmology is *general relativity* or the more rapidly evolving *Brans-Dicke theory* (Brans & Dicke, 1961[6]).

[6] Brans, C., & Dicke, R. H. (November, 1961). Mach's Principle and a Relativistic Theory of Gravitation. *Phys. Rev.*, 124, 925; https://doi.org/10.1103/PhysRev.124.925.

The above equilibrium argument may be applied also to the *neutrino* abundance. In the epoch where T > 10^{10} °K, the very high thermal *electron* and *photon* abundance would be sufficient to assure an equilibrium thermal abundance of *electron-type neutrinos*, assuming the presence of *neutrino-antineutrino* pair-production processes. This means that a strictly thermal *neutrino* and *antineutrino* distribution, in *thermal equilibrium* with the radiation, would have issued from the highly contracted phase. Conceivably, even gravitational radiation could be in *thermal equilibrium*.

Without some knowledge of the *density of matter* in the primordial fireball we cannot predict the present radiation temperature. However, a rough upper limit is provided by the observation that *black-body* radiation at a temperature of 40° K provides an *energy density* of 2 x 10^{-29} gm cm^3, very roughly the maximum *total energy density* compatible with the observed *Hubble constant* and *acceleration* parameter. Evidently, it would be of considerable interest to attempt to detect this primeval thermal radiation directly.

Two of us (P. G. R. and D. T. W.) have constructed a *radiometer* and receiving horn capable of an absolute measure of *thermal radiation* at a *wavelength* of 3 cm. The choice of *wavelength* was dictated by two considerations, that *at much shorter wavelengths atmospheric absorption would be troublesome*, while at *longer wavelengths galactic and extragalactic emission would be appreciable*. Extrapolating from the observed background radiation at longer *wavelengths* (~ 100 cm) *according to the power-law spectra characteristic of synchrotron radiation or bremsstrahlung*, we can conclude that the *total background* at 3 cm due to the Galaxy and the extragalactic sources should not exceed 5 x 10^{-3} °K when averaged over all directions. Radiation from *stars* at 3 cm is $< 10^{-9}$ °K. The contribution to the *background* due to the *atmosphere* is expected to be approximately 3.5° K, and this can be accurately measured by tipping the antenna (Dicke, Beringer, Kyhl, & Vane, 1946[7]).

[7] Dicke, R. H., Beringer, R., Kyhl, R. L., & Vane, A. B. (September, 1946). Atmospheric Absorption Measurements with a Microwave Radiometer. *Phys. Rev.*, 70, 340; https://doi.org/10.1103/PhysRev.70.340.

While we have not yet obtained results with our instrument, we recently learned that Penzias & Wilson (1965)[8] of the Bell Telephone Laboratories have observed *background radiation* at 7.3 cm *wavelength*.

[8] Penzias, A. A., & Wilson, R. W. (July, 1965). A Measurement of Excess Antenna Temperature at 4080 Mc/s. *Astrophys. J.*, 142, 1, 419–21; https://doi.org/10.1086/148307. See below.

In attempting to eliminate (or account for) every contribution to the noise seen at the output of their receiver, they ended with a residual of 3.5 ° ± 1°K. Apparently this could only be due to radiation of unknown origin entering the antenna.

It is evident that more measurements are needed to determine a spectrum, and we expect to continue our work at 3 cm. We also expect to go to a wavelength of 1 cm. We understand that measurements at wavelengths greater than 7 cm may be filled in by Penzias and Wilson.

A temperature in excess of 10^{10} ° K during the highly contracted phase of the universe is strongly implied by a present temperature of 3.5° K for *black-body* radiation. There are two reasonable cases to consider. Assuming a *singularity-free oscillating cosmology*, we believe that the temperature must have been high enough to decompose the heavy elements from the previous cycle, for there is no observational evidence for significant amounts of heavy elements in outer parts of the oldest stars in our Galaxy. *If the cosmological solution has a singularity*, the temperature would rise much higher than 10^{10} °K in approaching the singularity (see, e.g., Fig. 1).

It has been pointed out by one of us (P. J. E. P.) that the observation of a temperature as low as 3.5 °K, together with the estimated abundance of *helium* in the *protogalaxy*, provides some important evidence on possible cosmologies (Peebles, 1965)[9].

[9] Peebles, P J. E. (1965). *Phys. Rev.*, (in press).

This comes about in the following way. Considering again the epoch T 10^{10} °K, we see that the presence of the thermal *electrons* and *neutrinos* would have assured nearly equal abundances of *neutrons* and *protons*. Once the temperature has fallen so low that *photodissociation* of *deuterium* is not too great, the *neutrons* and *protons* can combine to form *deuterium*, which in turn readily burns to *helium*. This was the type of process envisioned by Gamow, Alpher, Herman, and others (Alpher,

Bethe, & Gamow, 1948[10]; Alpher, Follin, & Herman, 1953[11]; TIoyle & Tayler, 1964[12]).

[10] Alpher, R. A., Bethe, H. A., & Gamow, G. (April, 1948). The Origin of Chemical Elements. *Phys. Rev.*, 73, 803; https://doi.org/ 10.1103/PhysRev.73.803.

[11] Alpher, R. A., Follín, J. W., & Herman, R. C. (December, 1953). Physical Conditions in the Initial Stages of the Expanding Universe. *Phys. Rev.*, 92, 1347; https://doi.org/10.1103/PhysRev.92.1347.

[12] Hoyle, F., & Tayler, R. J. (September, 1964). The Mystery of the Cosmic Helium Abundance. *Nature*, 203, 1108-10; https://doi.org/ 10.1038/2031108a0.

Evidently the amount of *helium* produced depends on the *density of matter* at the time *helium* formation became possible. If at this time the *nucleon density* were great enough, an appreciable amount of *helium* would have been produced before the *density* fell too low for reactions to occur. Thus, from an upper limit on the possible *helium* abundance in the protogalaxy we can place an upper limit on the *matter density* at the time of *helium* formation (which occurs at a fairly definite temperature, almost independent of *density*) and hence, given the *density of matter* in the present universe, we have a lower limit on the present radiation temperature. This limit varies as the cube root of the assumed present mean *density of matter*.

While little is reliably known about the possible *helium* content of the *protogalaxy*, a reasonable upper bound consistent with present abundance observations is 25 per cent *helium* by *mass*. With this limit, and *assuming that general relativity is valid*, then if the present *radiation temperature* were 3.5° K, we conclude that the *matter density* in the universe could not exceed 3 x 10^{-32} gm cm^3. (See Peebles, 1965[9] for a detailed development of the factors determining this value.) This is a factor of 20 below the estimated average *density from matter* in galaxies (Oort, 1958[13]), but the estimate probably is not reliable enough to rule out this low *density*.

[13] Oort, J. H. (1958). *La Structure et l'evolution de l'universe*. 11th Solvay Conf., Éditions Stoops, Brussels, p. 163.

CONCLUSIONS

While all the data are not yet in hand, we propose to present here the possible conclusions to be drawn if we tentatively assume that the measurements of Penzias & Wilson (1965)[8] do indicate *black-body* radiation at 3.5° K. We also assume that the universe can be considered to be *isotropic* and *uniform*, and that the present *energy density* in *gravitational radiation* is a small part of the whole. Wheeler (1958)[4] has remarked that *gravitational radiation* could be important. For the purpose of obtaining definite numerical results, we take the present *Hubble redshift age* to be 10^{10} years. *Assuming the validity of Einstein's field equations*, the above discussion and numerical values impose severe restrictions on the cosmological problem. The possible conclusions are conveniently discussed under two headings, the assumption of a universe with either an open or a closed space.

Open universe. — From the present observations we cannot exclude the possibility that the total *density of matter* in the universe is substantially below the minimum value 2×10^{-29} gm cm^3 required for a *closed universe*. *Assuming general relativity is valid*, we have concluded from the discussion of the connection between *helium* production and the present radiation temperature that the present *density* of material in the universe must be $<\sim 3 \times 10^{-32}$ gm cm^3, a factor of 600 smaller than the limit for a *closed universe*. The thermal-radiation energy density is even smaller, and from the above arguments we expect the same to be true of neutrinos.

Apparently, *with the assumption of general relativity and a primordial temperature consistent with the present 3.5° K, we are forced to adopt an open space, with very low density. This rules out the possibility of an oscillating universe.* Furthermore, as Einstein (1950)[14] remarked, this result is distinctly non-Machian, in the sense that, with such a low *mass density*, we cannot reasonably assume that the local inertial properties of space are determined by the presence of *matter*, rather than by some absolute property of space.

[14] Einstein, A, (1950, *The Meaning of Relativity*, 3rd ed., Princeton University Press, Princeton, N.J., p. 107.

Closed universe. — This could be the type of *oscillating universe* visualized in the introductory remarks, *or it could be a universe expanding from a singular state.* In the framework of the present discussion the required *mass density* in excess of 2×10^{-29} gm cm^3 could not be due to *thermal radiation*, or to *neutrinos*, and it must be presumed that it is due to *ordinary matter*, perhaps *intergalactic gas* uniformly distributed or else in large clouds (small *protogalaxies*) that have not yet generated *stars* (see Fig. 1).

With this large *matter* content, the limit placed on the *radiation temperature* by the low *helium* content of the solar system is very severe. The present *black-body* temperature would be expected to exceed 30° K (Peebles, 1965). One way that we have found reasonably capable of decreasing this lower bound to 3.5° K is to introduce a *zero-mass* scalar field into the cosmology. *It is convenient to do this without invalidating the Einstein field equation*, and the form of the theory for which the scalar interaction appears as an *ordinary matter* interaction (Dicke, 1962[15]) has been employed.

[15] Dicke, R. H. (March, 1962). Mach's Principle and Invariance under Transformation of Units. *Phys. Rev.*, 125, 2163; https://doi.org/10.1103/PhysRev.125.2163.

The cosmological equation (Brans & Dicke, 1961[6]) was originally integrated for a *cold universe* only, but a recent investigation of the solutions for a *hot universe* indicates that with the scalar field the universe would have expanded through the *temperature* range $T \sim 10^9$ °K so fast that essentially no *helium* would have been formed. The reason for this is that the static part of the scalar field contributes a pressure just equal to the scalar-field *energy density*. By contrast, the pressure due to incoherent *electromagnetic radiation* or to *relativistic particles* is one third of the *energy density*. Thus, if we traced back to a highly contracted universe, we would find that the scalar-field *energy density* exceeded all other contributions, and that this fast-increasing scalar-field *energy* caused the universe to expand through the highly contracted phase much more rapidly than would be the case if the scalar field vanished. The essential element is that the pressure approaches the *energy density*, rather than one third of the *energy density*. Any other interaction which would cause this, such as the model given by Zel'dovich (1962)[16], would also prevent appreciable *helium* production in the highly contracted universe.

[16] Zel'dovich, Ya. B. (1962). *Soviet Phys.— J. E. T. P.*, 14, 1143.

Returning to the problem stated in the first paragraph, we conclude that it is possible to save *baryon* conservation in a reasonable way *if the universe is closed and oscillating*. To avoid a catastrophic *helium* production, either the present *matter density* should be $< 3 \times 10^{-32}$ gm/cm^3, or there should exist some form of *energy* content with very high pressure, such as the zero-mass scalar, capable of speeding the universe through the period of *helium* formation. To have a closed space, an *energy density* of 2×10^{-29} gm cm^3 is needed. Without a *zero-mass scalar*, or some other "hard" interaction, the *energy* could not be in the form of *ordinary matter* and may be presumed to be *gravitational radiation* (Wheeler, 1958[4]).

One other possibility for *closing the universe*, with *matter* providing the *energy content* of the universe, is the assumption that the universe contains a net *electron-type neutrino* abundance (in excess of *antineutrinos*) greatly larger than the *nucleon* abundance. In this case, if the *neutrino* abundance were so great that these *neutrinos* are degenerate, the degeneracy would have forced a negligible equilibrium *neutron* abundance in the early, highly contracted universe, *thus removing the possibility of nuclear reactions leading to helium formation.* However, the required ratio of *lepton* to *baryon number* must be $> 10^9$.

Arno Allan Penzias (1933–2024): A visionary explorer of the Universe.

K. I. Kellermann, National Radio Astronomy Observatory, Charlottesville, VA; April 30, 2024. PNAS Retrospective. *PNAS*; https://doi.org/10.1073/pnas.2405969121.

Arno Penzias together with Robert Wilson detected the faint three-degree cosmic microwave radiation from the big-bang creation of the Universe starting a revolution in cosmology.[1]

> [1] Penzias, A. A., & Wilson, R. W. (July, 1965). A Measurement of Excess Antenna Temperature at 4080 Mc/s. *Astrophys. J.,* 142, 1, 419–21; see below.

Fifteen years later, Penzias and Wilson joined by Keith Jefferts made the first detection of the millimeter-wave radiation from interstellar carbon monoxide, thus opening another new field of research—radio spectroscopy and interstellar chemistry.[2]

> [2] Wilson, R. W., Jefferts, K., & Penzias, A. A. (1970). Carbon monoxide in the Orion Nebula. *Astrophys. J.,* 161, L43–44.

Penzias was born in Munich Germany on April 26, 1933, and died in San Francisco, California at the age of 90 from complications of Alzheimer's disease on January 22, 2024. When only six years old, after narrowly avoiding deportation to a camp in Poland, Arno, together with his younger brother, Gunther, were sent alone on a train to England. A short time later, they were joined in the UK by their parents, and in 1939, the whole family emigrated to the United States. He married Anne Barras in 1954. They had three children, David, Mindy, and Laurie. Following their divorce, Arno married Sherry Levit in 1996.

After graduating from Brooklyn Technical High School, he received his BS in Physics from the City College of New York in 1954. Following two years in the Army Signal Corps developing radar systems, Arno went to Columbia University where he worked under Charles Townes to build a sensitive tunable maser amplifier as part of his dissertation research. Using the maser amplifier in an unsuccessful attempt to measure the 21-centimeter line emission from hydrogen in the Perseus cluster of galaxies, he received his PhD in physics from Columbia University in 1962.

Arno began his 37-year career at Bell Labs in 1961. Teaming up with Bell Labs colleague Robert Wilson, Arno began a radio astronomy program using the 20-foot horn reflector antenna, that was no longer needed for testing the AT&T satellite-linked intercontinental telephone network. He built a reference load cooled by liquid helium to 4.2 K that was compared with the output of the horn antenna by rapidly switching between the horn and the reference load using a switch built by Wilson[3].

[3] Penzias, A. A. (1965). Helium cooled reference noise source in a 4-kMc waveguide. *Rev. Sci. Instrum.*, 36, 68.

However, their results in May 1964 were perplexing. The horn was warmer than the 4.2 K degrees load although it should have been colder corresponding to the 2.3 K degrees expected from the atmosphere. Careful experiments confirmed that the contribution from the atmosphere was only 2.3 ± 0.3 K and that the noise due to losses in the antenna system was expected to be less than 0.9 degrees. Penzias and Wilson considered that the excess noise in their antenna might be interference from the heavily populated New York City only 20 miles away, but the additional noise was independent of antenna orientation. There was no dependence on the time of the day or the time of the year, thus eliminating any solar system or Galactic effect. They ruled out any significant contribution of the artificial radiation belts formed by high altitude thermonuclear tests conducted a few years earlier as the noise level remained constant as the radiation belts decayed with time. They also suspected that pigeon droppings inside the horn might be contributing to their anomalous results. They "evicted" the pigeons and cleaned out their droppings, but this too gave no improvement.

When Arno became aware of a planned experiment at nearby Princeton University to detect microwave radio noise originating from the early Universe, he and Wilson finally understood the source of the excess noise. The very short letter by Penzias and Wilson published in the Astrophysical Journal, was one of the most highly cited astronomy papers of the 20th century (1). Their discovery of the cosmic background radiation at the Bell Telephone Laboratories, for which they received the 1978 Nobel Prize in Physics, started a new field of precision cosmology.

Despite an early unsuccessful attempt to detect interstellar hydroxyl ions (OH), Arno remained excited about the prospects of studying interstellar molecules by means of their microwave transitions, which occur primarily at millimeter wavelengths. He proposed searching for interstellar molecular transitions near 3 millimeters wavelength using the NRAO 36-foot millimeter-wave radio telescope located on Kitt Peak near Tucson, Arizona. Working with Wilson and Keith Jefferts, a Bell Labs atomic physicist, Arno and his colleagues found an unexpectedly strong signal from carbon monoxide (CO) in the direction of the Orion Nebula (2). This opened the door for the millimeter-wave observations of a wide variety of interstellar molecules which became the hottest topic in radio astronomy.

Between times spent studying the cosmic microwave background and interstellar molecules, Penzias worked with Wilson on problems more directly concerning the telephone company including the atmospheric attenuation of microwave and infrared transmissions[4].

Starting with his 1972 appointment as Head of the Radio Physics Research Department, Arno steadily rose through the AT&T Bell Labs management ranks. In 1976, he became the Director of the Radio Research Laboratory; three years later, the Executive Director of Research and Communication; and in 1984, Chief Scientist and Vice President for Research. After retiring from Bell Labs in 1998, Arno moved to California where he began a new career as a Venture Partner at New Enterprise Associates.

In addition to being elected to the NAS in 1975 and … [sharing a half share of] the 1978 Nobel Physics Prize [with Wilson] ["for their discovery of cosmic microwave background radiation"], Arno was recognized by many other awards and recognitions including (with Robert Wilson) the NAS Henry Draper Medal in astronomical physics "for their discovery of the cosmic microwave radiation (a remnant of the very early universe), and their leading role in the discovery of interstellar molecules." He received honorary doctorates from more than 15 universities in the United States and abroad, held several patents, served on multiple editorial boards; and for more than 20 years, was the Vice President of the Committee of Concerned Scientists which he helped to found. In 1989, he authored the remarkably prescient book Ideas and Information: Managing in a High-Tech World about the future societal impact of computers[5].

[5] Penzias, A. A. (1989). *Ideas and Information: Managing in a High-Tech World*. Norton, New York.

He will be greatly missed by his many friends and colleagues from both academia and industry.

Penzias, A. A., & Wilson, R. W. (July, 1965). A Measurement of Excess Antenna Temperature at 4080 Mc/s.

Astrophys. J., 142, 1, 419–21; https://doi.org/10.1086/148307; also at https://articles.adsabs.harvard.edu/pdf/1965ApJ...142..419P.

Submitted May 13, 1965.

Bell Telephone Laboratories, Inc., Crawford Hill, Holmdel, New Jersey.

––––––––––––––

Measurements of the effective *zenith noise temperature* of the 20-foot horn-reflector antenna (Crawford, Hogg, & Hunt, 1961)[1] at the Crawford Hill Laboratory, Holmdel, New Jersey, at 4080 Mc/s have yielded a value about 3.5° K higher than expected.

> [1] Crawford, A B., Hogg, D. C., & Hunt, L. E. (July, 1961). Project Echo: A Horn-Reflector Antenna for Space Communication. *Bell System Tech. J.,* 40, 1095-116.

This excess *temperature* is, within the limits of our observations, isotropic, unpolarized, and free from seasonal variations (July, 1964-April, 1965). A possible explanation for the observed excess *noise temperature* is the one given by Dicke, Peebles, Roll, & Wilkinson (1965)[2] in a companion letter in this issue.

> [2] Dicke, R. H., Peebles, P. J. E., Roll, P. G., & Wilkinson, D. T. (July, 1965). Cosmic Black-body Radiation. *Astrophys. J.,* 142, 414. See above.

The *total antenna temperature* measured at the zenith is 6.7° K of which 2.3° K is due to atmospheric absorption. The calculated contribution due to ohmic losses in the antenna and back-lobe response is 0.9° K.

The radiometer used in this investigation has been described elsewhere. It employs a traveling-wave *maser*, a low-loss (0.027-db) comparison switch, and a liquid helium-cooled reference termination (Penzias, 1965)[4].

[4] Penzias, A. A. (1965). Helium cooled reference noise source in a 4-kMc waveguide. *Rev. Sei. Instr.*, 36, 68.

Measurements were made by switching manually between the antenna input and the reference termination. The antenna, reference termination, and radiometer were well matched so that a round-trip return loss of more than 55 db existed throughout the measurement; thus, errors in the measurement of the *effective temperature* due to impedance mismatch can be neglected. The estimated error in the measured value of the *total antenna temperature* is 0.3° K and comes largely from uncertainty in the absolute calibration of the reference termination.

The contribution to the *antenna temperature* due to atmospheric absorption was obtained by recording the variation in antenna temperature with elevation angle and employing the secant law. The result, $2.3° \pm 0.3°$ K, is in good agreement with published values (Hogg, 1959[5]; DeGrasse, Hogg, Ohm, & Scovil, 1959[6]; Ohm, 1961[7]).

[5] Hogg, D. C. (September, 1959). Effective Antenna Temperatures Due to Oxygen and Water Vapor in the Atmosphere. *J. Appl. Phys.*, 30, 1417-9; https://doi.org/10.1063/1.1735345.
[6] DeGrasse, R. W., Hogg, D. C., Ohm, E. A., & Scovil, H. E. D. (1959). Ultra-low Noise Receiving System for Satellite or Space Communication. *Proceedings of the National Electronics Conference,* 15, 370.
[7] Ohm, E. A. (July, 1961). Project Echo: Receiving System. *Bell System Tech. J.*, 40, 1065.

The contribution to the *antenna temperature* from ohmic losses is computed to be $0.8° \pm 0.4°$ K. In this calculation we have divided the antenna into three parts: (1) two non-uniform tapers approximately 1 m in total length which transform between the 2 1/8-inch round output waveguide and the 6-inch-square antenna throat opening; (2) a double-choke rotary joint located between these two tapers; (3) the antenna itself. Care was taken to clean and align joints between these parts so that they would not significantly increase the loss in the structure. Appropriate tests were made for leakage and loss in the rotary joint with negative results.

The possibility of losses in the *antenna horn* due to imperfections in its seams was eliminated by means of a taping test. Taping all the seams in the section near the throat and most of the others with aluminum tape caused no observable change in antenna temperature.

The backlobe response to ground radiation is taken to be less than $0.1°$ K for two reasons: (1) Measurements of the response of the antenna to a small transmitter located on the ground in its vicinity indicate that the average back-lobe level is more than 30 db below isotropic response. The horn-reflector antenna was pointed to the zenith for these measurements, and complete rotations in azimuth were made with the transmitter in each of ten locations using horizontal and vertical transmitted polarization from each position. (2) Measurements on smaller horn-reflector antennas at these laboratories, using pulsed measuring sets on flat antenna ranges, have consistently shown a back-lobe level of 30 db below isotropic response. Our larger antenna would be expected to have an even lower back-lobe level.

From a combination of the above, we compute the remaining unaccounted-for *antenna temperature* to be $3.5° \pm 1.0°$ K at 4080 Mc/s. In connection with this result, it should be noted that DeGrasse *el al.* (1959)[2] and Ohm (1961)[7] give total system temperatures at 5650 Mc/s and 2390 Mc/s, respectively. From these it is possible to infer upper limits to the *background temperatures* at these frequencies. These limits are, in both cases, of the same general magnitude as our value.

Note added in proof. — The highest *frequency* at which the *background temperature* of the sky had been measured previously was 404 Mc/s (Pauliny-Toth & Shakeshaft, 1962)[8], where a *minimum temperature* of $16°$ K was observed.

[8] Pauliny-Toth, I. I. K., & Shakeshaft, J. R. (1962). A survey of the background radiation at a frequency of 404 Mc/s, I. *MNRAS*, 124, 61; https://doi.org/10.1093/mnras/124.1.61.

Combining this value with our result, we find that the *average spectrum* of the *background radiation* over this *frequency* range can be no steeper than λ^{07}. This clearly eliminates the possibility that the radiation we observe is due to radio sources of types known to exist, since in this event, the spectrum would have to be very much steeper.

Hannes Alfvén - Nobel Lecture, December 11, 1970. Plasma physics, space research and the origin of the solar system.

Hannes Alfvén – Nobel Lecture. NobelPrize.org. https://www.nobelprize.org/uploads/2018/06/alfven-lecture.pdf.

[The Nobel Prize in Physics 1970 was divided equally between Hannes Olof Gösta Alfvén "for fundamental work and discoveries in *magnetohydro-dynamics* with fruitful applications in different parts of *plasma physics*" and Louis Eugène Félix Néel "for fundamental work and discoveries concerning antiferromagnetism and ferrimagnetism which have led to important applications in solid state physics". The Nobel Prize in Physics 1970. NobelPrize.org. https://www.nobelprize.org/prizes/physics/1970/summary/.

The phenomenon of *aurora borealis* occurs when bursts of charged particles from the sun collide with the earth's *magnetic field*. These *jets* of particles are an example of a special state of matter—*plasma. Plasma* is a gas comprised of electrons and ions (electrically charged atoms) that forms at high temperatures. From the late 1930s onward, Hannes Alfvén developed a theory about *aurora borealis*, which led to *magneto-hydrodynamics*; the theory of the relationships between a *plasma's* movements, *electric currents and fields*, and *magnetic fields*. Hannes Alfvén – Facts. NobelPrize.org. <https://www.nobelprize.org/prizes/physics/1970/alfven/facts/>

---

1. *Science and instruments*

The center of gravity of the physical sciences is always moving. Every new discovery displaces the interest and the emphasis. Equally important is that new technological developments open new fields for

90

scientific investigation. To a considerable extent the way science takes depends on the construction of new instruments as is evident from the history of science. For example, after the development of classical mechanics and electromagnetism during the 19th century, a new era was started by the construction of highly developed spectrographs in the beginning of this century. For its time those were very complicated and expensive instruments. They made possible the exploration of the outer regions of the atom. Similarly, in the thirties the cyclotron - for its time a very complicated and expensive instrument - was of major importance in the exploration of the nucleus. Finally, the last decade has witnessed the construction of still more complicated and expensive instruments, the space vehicles, which are launched by a highly developed rocket technology and instrumented with the most sophisticated electronic devices. We may then ask the question: What new fields of research - if any - do these open for scientific investigation? Is it true, also in this case, that the center of gravity of physics moves with the big instruments?

2. Scientific aims of space research

The first decade of space research mainly concentrated on the exploration of space near the Earth: the *magnetosphere* and *interplanetary space*. These regions earlier were supposed to be void and structureless but we now know that they are filled with *plasmas*, intersected by sheath like *discontinuities*, and permeated by a complicated pattern of *electric currents* and *electric and magnetic fields*. The knowledge gained in this way is fundamental to our general understanding of *plasmas*, especially *cosmic plasmas*. Indirectly it will hence be important to *thermonuclear research*, to the study of the *structure of the galaxy and the metagalaxy*, and to *cosmological problems*. Our advancing knowledge in *cosmical electrodynamics* will make it possible to approach these fields in a less speculative way than hitherto. The knowledge of *plasmas* is also fundamental to our understanding of the *origin and evolution of the Solar System*, because there are good reasons to believe that the matter which now forms the celestial bodies once was dispersed in a *plasma* state.

The second decade of space research seems to display a different character, at least to some extent. As several of the basic problems of the *magnetosphere* and *interplanetary space* are still unsolved, one can be sure that these regions will still command much interest. However, the lunar landings and also the deep-space probes to Venus and Mars have supplied us with so many new scientific facts that the emphasis in space research is moving towards the exploration of the Moon, the planets, and other celestial bodies in the *Solar System*.

The first phase of this exploration is necessarily of a character somewhat similar to the exploration of the polar regions and other regions of the earth which have been difficult to reach: a detailed mapping-out combined with geological, seismic, magnetic, and gravity surveys and an exploration of the atmospheric conditions. However, when applying this research pattern to the *Moon* and the *planets* one is confronted with another problem, viz. *how these bodies were originally formed*. In fact, many of the recent space research reports end with speculations about the *formation and evolution of the solar system*. It seems that this will necessarily be one of the main problems - perhaps the main problem - on which space research will center in the near future. Already at an early data NASA stated that the main scientific goal of space research should be to clarify how the solar system was formed. This is indeed one of the fundamental problems of science. We are trying to write the scientific version of *how our Earth and its neighbors once were created*. From a - shall we say - philosophical point of view, this is just as important as the *structure of matter*, which has absorbed most of the interest during the first two thirds of this century.

3. *Plasma physics and its applications*

Before we concentrate on our main topic: *how the solar system originated*, we should make a brief summary of the state of *plasma physics*. As you know, *plasma physics* has started along two parallel lines. *The first one* was the hundred years old investigations in what was called *electrical discharges in gases*. This approach was to a high degree experimental and phenomenological, and only very slowly reached some degree of theoretical sophistication. Most theoretical physicists looked down on this field, which was complicated and awkward. *The plasma exhibited striations and double-layers, the electron distribution was non-Maxwellian, there were all sorts of oscillations and in stabilities*. In short, it was a field which was not at all suited for mathematically elegant theories.

The other approach came from the highly developed *kinetic theory of ordinary gases*. It was thought that with a limited amount of work this field could be extended to include also *ionized gases*. The theories were mathematically elegant and when drawing the consequences of them it was found that *it should be possible to produce a very hot plasma and confine it magnetically*. This was the starting point of *thermonuclear* research.

However, these theories had initially very little contact with experimental *plasma physics*, and all the awkward and complicated phenomena which had been treated in the study of *discharges in gases* were simply neglected. The result of this was what has been called the *thermonuclear crisis* some 10 years ago. It taught us that *plasma physics* is a very difficult field, which can only be developed by a close cooperation between theory and experiments. As H.S.W. Massey once said (in a somewhat different context): "The human brain alone is not able to work out the details and understanding of the inner workings of natural processes. Without laboratory experiment there would be no physical science today."

The *cosmical plasma physics* of today is far less advanced than the *thermonuclear research physics*. It is to some extent the playground of theoreticians who have never seen a *plasma* in a laboratory. Many of them still believe in formulae which we know from laboratory experiments to be wrong. The *astrophysical* correspondence to the *thermonuclear crisis* has not yet come.

I think it is evident now that in certain respects *the first approach to the physics of cosmical plasmas has been a failure*. It turns out that in several important cases this approach has not given even a first approximation to truth but led into dead-end streets from which we now have to turn back. The reason for this is that *several of the basic concepts on which the theories are founded, are not applicable to the condition prevailing in cosmos*. They are "generally accepted" by most theoreticians, they are developed with the most sophisticated mathematical methods and *it is only the plasma itself which does not "understand", how beautiful the theories are and absolutely refuses to obey them*. It is now obvious that we have to start a *second approach* from widely different starting points.

4. *Characteristics of first and second approach to cosmic plasma physics*

The two different approaches can be summarized in Table 1.

If you ask where the border goes between the *first approach* and the *second approach* today, an approximate answer is that it is given by the reach of spacecrafts. This means that in every region where it is possible to explore the state of the *plasma* by *magnetometers, electric field probes* and *particle analyzers*, we find that *in spite of all their elegance, the first approach theories have very little to do with reality*. It seems that the change from the *first approach* to the *second approach* is the astrophysical correspondence to the thermonuclear crisis.

Table 1 Cosmical electrodynamics

First approach	*Second approach*
Homogeneous models	Space *plasmas* have often a complicated inhomogeneous structure.
Conductivity $\sigma = \text{co}$ Electric field $E_g = 0$	Depends on current and often suddenly becomes 0, E_g often $\neq 0$.

Magnetic field lines are "frozen-in" and "move" with the *plasma*.	Frozen-in picture often completely misleading.
Electromagnetic conditions illustrated by magnetic field line picture.	It is equally important to draw the current lines and discuss the electric circuit.
Electrostatic double layers neglected.	Electrostatic double layers are of decisive importance in low density *plasmas*.
Filamentary structures and current sheets neglected or treated inadequately.	Currents produce filaments or flow in thin sheets.
Theories mathematically elegant and very well developed.	Theories still not very well developed and partly phenomenological.

5. *The origin of the solar system*

From what has been said it is obvious that astrophysics runs the risk of getting too speculative, unless it tries very hard to keep contact with laboratory physics. Indeed, it is essential to stress that *astrophysics is essentially an application to cosmic phenomena of the laws of nature found in the laboratory.* From this follows that a particular field of astrophysics is not ripe for a scientific approach before experimental physics has reached a certain state of development. As a well-known historic example, before the advance of nuclear physics the attempts to understand how the stars generated their energy could not possibly be more than speculations without very much permanent value.

The problem of how the *solar system* originated has been the subject of a large number of highly divergent hypotheses. The reason for this has been that there was not enough basic knowledge of physics in the fields essential for the understanding of the phenomena and for a decision about which processes were possible.

However, before we discuss any details of a *theory of the origin and evolution of the solar system,* it is essential to define what general character such a theory should have. In the past *too much attention has been concentrated on the formation of planets around the sun.* One of the unfortunate results of this is that many theories of the origin of the solar system have been based on theories of the early history of the sun. This is a very shaky basis because *the formation of the sun (and other stars) is a highly controversial subject.* Recognizing that the satellite systems of Jupiter, Saturn, and Uranus are very similar to the planetary system, and at least as regular as this system, it seems now more appropriate to aim at a general theory of the formation of secondary bodies around a central body, regarding the formation of the planetary system as only one of the applications of such a general theory.

The study of the sequence of processes by which the *solar system* originated has often been called *cosmogony,* a term which, however, is used in many other connections. As the *origin of the solar system* is essentially a question of the repeated formation of secondary bodies around a primary body, the term *hetegony* (from Greek *hetairos* or *hetes* = companion) has been suggested.

$$\ldots$$

Fig. 1

It seems likely (and is fairly generally agreed) that the sequence of events leading to the formation of the solar system is likely to have been as shown in Fig. 1 (we are here following what has been called the "planetesimal" approach). *A primeval plasma was concentrated in certain regions around a central body, and condensed to small solid grains.* (Even the primeval *plasma* may have contained grains.) The *grains* accreted to what have been called *embryos* and by further accretion larger bodies were formed: *planets* if the central body was the *sun,* and *satellites* if it was a *planet.* The place of the *asteroids* in the *hetegonic* diagram is controversial. They have formerly been generally considered to be fragments of a broken-up *planet,* but there are now an increasing number of arguments for the view that they represent - or at least are similar to - an intermediate state in the formation of planets. A clarification of these two alternatives is important.

Even if the diagram of Fig. 1 is fairly generally accepted as it stands, this does not mean that the different processes are clarified. To a high degree they are still of a hypothetical character. Up to rather recently this has necessarily been the case because the basic processes have not been known very well. To some extent we have been in the same situation as the astrophysicists trying to clarify the energy generation in *stars* before the advent of nuclear physics. However, the situation seems now to be changing so that there is a good hope to bring the whole field of research from the state of a discussion of more or less bright hypotheses to a systematic scientific analysis.

6. *Basic knowledge for the reconstruction of the hetegonic processes*

Besides *plasma* physics, which we have already discussed, there are a number of other fields of research which are basic for the reconstruction of the *hetegonic* processes.

(1) *Plasma chemistry* means the field of research concerned with chemical reactions in a *plasma. These are basically different from the reactions in non-ionized gases.* It should also be considered to include *the separation of different elements which takes place in an inhomogeneous plasma due to, e.g. temperature gradients and electric currents.* Furthermore, *the interaction between a plasma and a solid grain condensing from it is highly dependent on the state of ionization.* The laboratory results and their application to cosmic conditions are relevant for the understanding of the different chemical composition of the celestial bodies.

For the next process in our evolutionary diagram, viz. the *accretion of larger bodies* from the initial condensation, the following fields of research are essential.

(2) *Solid-body collisions*. The *grains* which are the primary result of the condensation will move in Kepler orbits around the central body, but their motion will be disturbed by several effects. *One of them is due to the mutual collisions*. The relative velocities at these collisions may have any value from zero up to some 10 km/sec. This means that in many cases we are in the region of *"hypervelocity" collision*s. This is a field which is not yet understood very well. Available laboratory results seem to be scarce, and their application to cosmic conditions is uncertain because we know very little about the structure of the *grains*. Collisions between bodies with fluffy shock-absorbing surface layers are likely to differ from collisions between hard "marbles". *Meteorite* studies are supplying us with some information. The Apollo results about meteoroid impact on the lunar surface are another important source of knowledge. In these cases, however, *we do not gain very much information about the structure of the grains in space*, because the particles we recover have either passed the terrestrial atmosphere or been destroyed by impact on the lunar surface.

(3) The study *Kepler motion in a viscous medium* is essential for our understanding of the evolution of the orbits of the *grains* and the *embryos*. From a formal point of view this problem is similar to some basic problems in *plasma physics*, which are also concerned with a large number of interacting particles. It turns out that *in the neighborhood of a central body the condensed grains have a tendency to move in similar orbits, thus forming what have been called "jet streams" in space*.

(4) *Celestial mechanics* serves of course as a general background for the whole *hetegonic* process. This field has been rejuvenated by the application of computer analysis to many of the problems which were formerly impossible to handle. Connected with this is the discovery of the importance of *resonance* phenomena in the present structure of the *solar system*. It seems likely that at *hetegonic* times also *resonances* played a decisive role.

(5) The *hetegonic* processes took place 4 to 5 billion years ago. The evolution of the primary product of these processes into our present-day *solar system* has consisted of a number of relatively slow changes: *geological forces* have transformed the structure of the planets, *tidal effects* have braked the *spins* of some of the bodies (especially of the satellites), *collisions* have taken place in the *asteroidal belt*, and there have been *meteor impacts* on the planetary surfaces, etc. All these effects are important for the reconstruction of the state of the system immediately after the *hetegonic* processes ended. It is only after "correcting" for them that the solar system data we observe today are of value for the reconstruction of the *hetegonic* processes.

7. *Space observations relevant to the hetegonic problem*

From the analysis we have made it is evident that the background knowledge necessary for the understanding of the *hetegonic* processes is rapidly increasing through advances in several different fields of research. We shall now discuss the question of what sort of space missions are of special value for the study of the *hetegonic* problem.

Let us first state that many of the space missions which are carried out today or planned for the future give valuable contributions. Increased knowledge of the behavior of *cosmic plasmas* is gained by spacecraft carrying out *plasma* and *particle* measurements in the *magnetosphere* and *interplanetary space*. Further, *meteor* impacts on spacecraft supply us with information of the very small bodies in our environment, which are probably related to those small bodies out of which our present planets once accreted. Particularly important is the study of *meteor* impacts on the Moon (and on Mars). Hence these and other investigations "automatically" contribute to the background knowledge necessary for the solution of the *hetegonic* problem. But although this is satisfactory there are a number of crucial problems which cannot be solved unless space research is purposely directed towards solving them. We shall now discuss how this could be done.

8. *Big bodies versus small bodies*

It is usually thought that after the lunar landings the most important missions will be those to Venus, Mars, and the other planets. This is not necessarily true, because missions to *asteroids* and *comets* would be at least as interesting from a scientific point of view. As some asteroids are the closest neighbors of the *Earth-Moon system*, this would also be the easiest from a technical point of view.

Our analysis has indicated which fields of research will contribute to the clarification of different phases of the *hetegonic* processes. *Plasma physics* and *plasma chemistry* are important for the first phase, including the *condensation of small grains*. The study of *meteorite* and *asteroid* sized bodies will have bearing on the *accretion*. We can state as a general rule that *the smaller the body the further will the study of it bring us back in time*. Thus, small bodies will be relevant to earlier periods more than large bodies. This means that it is essentially through studying the properties of small bodies in space that we can hope to understand the crucial phase in the *formation of the solar system* when most of the *matter*, which later formed the *planets* and *satellites*, was still dispersed.

There is evidence that during the formation of the *planets* and *satellites* a great deal of information about the formation processes was stored in them. However, to a large extent this information is either obliterated or inaccessible. The *planets* are likely to have accreted from "*planetesimals*". The earliest phase of this *accretion* produced a small body, the *matter* of which may today be in the core of the *planet*, which means that it is inaccessible even if a manned spacecraft should land on the surface of the *planet*. There is also a possibility that, for example, *convection* in the interior of the planet has more or less completely obliterated the information once stored there. Concerning the surface layers, geological processes, including atmospheric effects, have mostly wiped out the surface traces of the *hetegonic* processes in Earth and probably also in Venus. In other bodies like the Moon and Mars, and

100

probably also Mercury, there seems to be considerable information left, but only referring to the very last phase of the *hetegonic* processes.

Hence our conclusion is that *studies of large bodies like the planets has only a limited value for the study of the origin of the solar system.*

Asteroids, comets and *meteoroids* are different in this respect. Even if some of these bodies are fragments produced at collisions in space it is very likely that also these fragments contain considerable information about the *condensation* and the *accretion* processes. Because of the smallness of the bodies there is no heating or convection in their interior which can obliterate the information stored from the time when they were produced, and at least in the very small bodies and by the fragmented bodies their "interior" is accessible. Furthermore, a study of them will give us knowledge of the behavior of small bodies in space which will be valuable for the clarification of the *hetegonic* processes in general. We study in them intermediate products in the manufacturing of *planets*. They give us, so-to-say, snapshots showing the sequence of events when a *planet* like the Earth once was created.

9. Old and new fields of science

We shall now return from our odyssey in both space and time to our starting point - how new technologies displace the center of gravity of the physical sciences. The great revolution in physics which took place in the beginning of this century meant that *classical mechanics* and *classical electrodynamics* were considered to be more or less obsolete as fields of research. The new fields which attracted the interest were the *theory of relativity* and *quantum mechanics* and the experimental work was largely concentrated on the exploration of the *electron shells of the atom*. The advance of *nuclear physics* marked another step in a similar direction.

The new trend which is introduced by the rise of *plasma physics* and *space research* is to some extent opposite. In these fields *quantum mechanics* and the *theory of relativity* are not very important. Instead, *classical mechanics* has become rejuvenated and is essential not only for calculating the trajectories of spacecraft but also for the study of the motion of the natural celestial bodies during their evolutionary history. Also, *classical electromagnetism* is of decisive importance to the *theory of magnetized plasmas*, which is basic both for *thermonuclear research* and for *astrophysics* in general. This does not mean that we should make the mistake - similar to what was made 50 years ago - of declaring the *atomic* and *nuclear physics* to be obsolete. They are not. They have an enormous inertia which will keep them moving, and they will produce many new and interesting results. But they have got very serious competitors, and remarkably enough these are the fields which earlier were declared dead that are now being resurrected.

It is possible that this new era also means a partial return to more understandable physics. For the non-specialists, *four-dimensional relativity theory*, and the *indeterminism of atom structure* have always been mystic and difficult to understand. I believe that it is easier to explain the *33 instabilities in plasma physics* or the *resonance structure of the solar system*. The increased emphasis on the new fields means a certain demystification of physics. In the spiral or trochoidal motion which science makes during the centuries, its guiding center has returned to those regions from where it started. It was the wonders of the night sky, observed by Indians, Sumerians or Egyptians, that started science several thousand years ago. It was the question why the wanderers - the planets - moved as they did that triggered off the scientific avalanche several hundred years ago. The same objects are now again in the center of science only the questions we ask are different. We now ask how to go there, and we also ask *how these bodies once were formed*. And if the night sky on which we observe them is at a high latitude, outside this lecture hall - perhaps over a small island in the archipelago of Stockholm - we may also see in the sky an *aurora*, which is a *cosmic plasma*, reminding us of the time when our world was born out of *plasma. Because in the beginning was the plasma.*

Eric J. Lerner (born May 31, 1947).

[*According to Expedia*,] Lerner is an American *popular science writer* [???] and *independent plasma* researcher. He wrote the 1991 book *The Big Bang Never Happened*, which advocates Hannes Alfvén's *plasma cosmology* instead of the *Big Bang* theory. He is founder, president, and chief scientist of LPP Fusion, advanced technology research, consulting and communications firm, which conducts scientific research in plasma physics, the development of an atomizing desalination process, and developing an advanced approach to economical fusion and new theories of quasars and cosmology. (See website; https://www.lppfusion.com/ eric-lerner/.)

[*According to Expedia*,] Lerner is an active *general science writer* [???], estimating that he has had about 600 articles published.

> [The description by Expedia, "*Popular science writer*" or "*general science writer*", is a derogatory term for a scientist who does not agree with conventional physics such as *Einstein's theories of relativity* and the *Big Bang theory*. Lerner is obviously a *highly accomplished expert in plasma physics*, despite not being a professor at a university. Between December, 1986, he published one book, one chapter in a book, and more than 11 highly technical articles in academic journals, on *cosmic plasma physics*, which were far beyond the comprehension of any popular audience.]

[*Publications by Lerner referred to in this volume*:

Lerner, E. J. (May, 1986). Magnetic self-compression in laboratory plasmas, quasars and radio galaxies. Part I. Laser and Particle Beams, 4, 193-213; https://doi.org/10.1017/S0263034600001750.

Lerner, E. J. (December, 1986). Magnetic Vortex Filaments, Universal Invariants and the Fundamental Constants. *IEEE Trans. Plasma Sci.*, Special Issue on Cosmic Plasma, PS-14, 6, 690-702; https://doi.org/10.1109/TPS.1986.4316620.

Lerner, E. J. (August, 1988). Plasma model of microwave background and primordial elements: an alternative to the big bang. *Laser and Particle Beams*, 6, 456-69; https://doi.org/10.1017/S0263034600005395.

Lerner, E. J. (April, 1989). Galactic Model of Element Formation. *IEEE Transactions on Plasma Science*, 17, 3, 259-63; https://doi.org/10.1109/27.24633.

Lerner, E. J. (September, 1990). Radio absorption by the intergalactic medium. *Astrophys. J.*, 361, 63-8; https://adsabs.harvard.edu/full/1990ApJ...361...63L.

Lerner, E. J. (October, 1992). [*Book*: 569 pages.] The Big Bang Never Happened—A Reassessment of the Galactic Origin of Light Elements (GOLE) Hypothesis and its Implications. Knopf Doubleday Publishing Group; https://www.lppfusion.com/storage/GOLE-Lerner.pdf.

Lerner, E. J. (December, 1992). Force-Free Magnetic Filaments and the Cosmic Background Radiation. *IEEE Trans. Plasma Sci.*, 20, 6, 935-8; https://ieeexplore.ieee.org/document/199554.

Lerner, E. J. (1993). [*Chapter*:] The Case Against the Big Bang. In: Arp, H. C., Keys, C. R., Rudnicki, K. (eds) Progress in New Cosmologies. Springer, Boston, MA; https://doi.org/10.1007/978-1-4899-1225-1_7.

Lerner, E. J. (September, 1993). Confirmation of radio absorption by the intergalactic medium. *Astrophys. & Space Sci.*, 207, 17-26; https://doi.org/10.1007/BF00659126.

Lerner, E. J. (May, 1995). Intergalactic Radio Absorption and the COBE Data. *Astrophys. Space Sci.*, 227, 1-2, 61–81; https://doi.org/10.1007/BF00678067.

Lerner, E. J. (July, 2008) Dr. Wright is Wrong-- a reply to Ned Wright's "Errors in The Big Bang Never Happened"; https://web.archive.org/web/20160108004949/http://bigbangneverhappened.org/p25.htm

Lerner, E. J., Falomo, R., & Scarpa, R. (2014). UV surface brightness of galaxies from the local universe to z ~ 5. *International Journal of Modern Physics* D, 23, 06, 1450058; https://doi.org/10.1142/S0218271814500588.

Lerner, E. J. (October, 2022). The Big Bang Never Happened—A Reassessment of the Galactic Origin of Light Elements (GOLE) Hypothesis and its Implications. https://www.lppfusion.com/storage/GOLE-Lerner.pdf.]

Lerner received a BA in physics from Columbia University and started as a graduate student in physics at the University of Maryland, but left after a year due to his dissatisfaction with the mathematical rather than experimental approach there. In 2006 he was a visiting scientist at the European Southern Observatory in Chile.

In his book, [E. J. Lerner (1991). The Big Bang Never Happened. New York and Toronto: Random House, pages 12 - 14, footnote on page 388, 286 - 316, 242] Lerner rejects mainstream *Big Bang cosmology*, and instead advances a non-standard *plasma cosmology* originally proposed by Hannes Alfvén, the 1970 Nobel Prize recipient in Physics. [See above.] The book appeared at a time when results from the Cosmic Background Explorer satellite were of some concern to astrophysicists who expected to see *cosmic microwave background* (CMB) *anisotropies* but instead measured a *blackbody* spectrum with little variation across the sky. Lerner referred to this as evidence that the *Big Bang* was a failed paradigm. He also denigrated the observational evidence for *dark matter* and recounted a well-known cosmological feature that *superclusters are larger than the largest structures that could have formed through gravitational collapse in the age of the universe.*

As an alternative to the *Big Bang*, Lerner adopted *Alfvén's model of plasma cosmology* that relied on *plasma physics* to explain most, if not all, cosmological observations. Lerner's ideas have been rejected by most *mainstream* physicists and cosmologists. These critics have claimed that, contrary to Lerner's assertions, *the size of superclusters is consistent with having arisen from a power spectrum of density fluctuations growing from the quantum fluctuations predicted in inflationary models.*

Physical cosmologists who have commented on the book have generally dismissed it. In particular, American astrophysicist and cosmologist Edward L. Wright criticized Lerner for making errors of fact and interpretation, arguing that:

- Lerner's alternative model for Hubble's law is *dynamically unstable*
- the *number density of distant radio sources* falsifies Lerner's explanation for the *cosmic microwave background*
- Lerner's explanation that *the helium abundance is due to stellar nucleosynthesis* fails because of the small observed abundance of heavier elements.

(See below.)

Lerner has disputed Wright's critique. (See below.)

Since 1994 LPP Fusion has been carrying out research on fusion and *fusion propulsion* funded by NASA through Jet Propulsion Laboratory, using *laboratory plasma fusion devices*, performing experimental work on a machine called a *dense plasma focus* (DPF). *NASA's Jet Propulsion Laboratory has funded mainstream as well as alternative approaches* [???] *to fusion*, and between 1994 and 2001 NASA provided a grant to *Lawrenceville Plasma Physics* (LPP), the company of which Lerner was the only employee, to explore whether Lerner's alternative approach to *fusion* might be useful to propel spacecraft; a 2007 *New York Times* article noted that Lerner had not received funding from the US Department of Energy.

Lerner believes that a *dense plasma focus* can also be used to produce useful *aneutronic fusion energy*. On November 14, 2008, Lerner received funding for continued research, to test the scientific feasibility of Focus Fusion. On January 28, 2011, LPP published preliminary results. In March 2012, the company published a paper saying that it had achieved temperatures of 1.8 billion degrees, beating the old record of 1.1 billion that had survived since 1978. In 2017, Lerner, *et al.* published evidence of confined ion energies in excess of 200 keV, with the best "shot" having a mean ion energy of 240 keV ± 20 keV which was reported as a record for confined fusion *plasmas*.

In October 2021, the company announced improved results with the latest version of its device, with reduced erosion and higher temperatures. In March, 2023 Lerner, *et al.* published a paper in a special issue of the *Journal of Fusion Energy*, showing that LPPFusion led all fusion companies in achieving" the highest ratio of fusion energy generation to device energy input. [Lerner, E. J., Hassan, S. M., Karamitsos-Zivkovic, I. *et al.* (March, 2023). Focus Fusion: Overview of Progress Towards p-B[11] Fusion with the Dense Plasma Focus. *J. Fusion Energ.*, 42, 7; https://doi.org/10.1007/s10894-023-00345-z.]

Lerner, E. J. (August, 1988). Plasma model of microwave background and primordial elements: an alternative to the big bang*.

Lasers and Particle Beams, 6, 457-69; https://doi.org/10.1017/ S0263034600005395.

* Paper dedicated to Professor Hannes Alfvén on the occasion of his 80th birthday, May 30, 1988.

Received October 1, 1987.
Accepted November 15, 1987.

Abstract

A plasma model of the origin of the light elements and the microwave background is presented. In contrast to the conventional Big Bang hypothesis, *the model assumes that helium, deuterium and the microwave background were all generated by massive stars in the early stages of galaxy formation.* The *microwave background* is scattered and isotropized by multi-GeV *electrons* trapped in the jets emitted by *active galactic nuclei.* The model produces reasonable amounts of heavy elements, *accurately predicts the gamma-ray background intensity and spectrum,* and explains the statistics of *quasars, compact* and *extended radio sources.*

---

1. Introduction

The supporters of the *Big Bang theory* have long pointed to three phenomena as evidence for this theory: the *isotropic microwave background,* the *universal abundance of helium,* and the *universal abundance of deuterium* (Wilkinson, 1986[1]).

108

[1] Wilkinson, D. T. (June, 1986). Anisotropy of the cosmic blackbody radiation. *Science*, 232, 4757, 1517-22; https://doi.org/10.1126/science.232.4757.1517.

In fact, *all three phenomena can be accounted for by an entirely different physical process: thermonuclear burning by massive stars in early galaxies.* The [*cosmic*] *microwave background* (CMB) is the *thermalized and isotropized radiation from such stars,* which simultaneously produced the primordial helium and deuterium.

This counterhypothesis has been advanced repeatedly over the past decades (e.g. Rees 1978[2]).

[2] Rees, M. J. (September, 1978). Origin of pregalactic microwave background. *Nature*, 275, 35-7; https://doi.org/10.1038/275035a0.

However, two primary objections have been raised against it. First, *some have argued that intergalactic space is transparent to microwave radiation,* and thus such radiation could not have been isotropized after its emission from galaxies. Second, *others have contended that the production of helium by thermonuclear processes would inevitably lead to more heavy elements,* such as carbon, oxygen and iron, than have actually been observed.

I show in this paper that, by correctly incorporating plasma processes, both these objections are overcome. The microwave background is scattered and isotropized by multi-Gev electrons trapped in jets emitted by active galactic nuclei. This scattering occurs through *synchrotron absorption and reemission.* The *over production of heavy elements does not occur because the stars most favored by early galactic formation processes are of intermediate mass and diffuse to the interstellar medium* almost exclusively helium.

The theory proposed here not only accounts for the phenomena cited as evidence for the *Big Bang*, but as well provides *excellent agreement with the observed gamma ray background* and *correctly predicts many features of radio source distribution and size.*

2. Temperature of the Cosmic Microwave Background Radiation and Energy Released from Helium

We begin by equating the energy density of the [*cosmic*] *microwave background* (CMB) with the *energy released in the generation of the observed helium,*

$$n_1 E_h H_e = 7.59 \times 10^{-15} T^4 \tag{1}$$

where n_1 is the *local cosmic density* (cm^{-3}), E_h is the *energy release* per H fused to H_e, H_e is the '*primordial*' ratio of He/H by *mass*, and T is the *microwave radiation temperature* (K). If we take $H_e = 0.24$, and since $E_h = 1.17 \times 10^{-15}$ erg, we get

$$T = 1.44 \times 10^2 \, n_1^{\frac{1}{4}}. \tag{2}$$

Recent observations by Tully (1986)[3] show that the *local matter density* is about twice the average in the larger region of *concentration* that extends about 200 Mpc from earth.

[3] Tully, R. B. (April, 1986). Alignment of clusters and galaxies on scales up to 0.1 C. *Astrophys. J.*, 303, 25-38.

Since recent observations and simulations (Valtonen & Byrd, 1986[4]; Tyson, 1984[5]) *have eliminated the observational basis for 'dark matter'* the *average density* within this large region can be calculated to be around $8 \times 10^{-8}/cm^3$, so n_1 is $1.6 \times 10^{-7}/cm^3$ (Lerner, 1986)[6].

[4] Valtonen, M. J. & Byrd, G. G. (April, 1986). Redshift asymmetries in systems of galaxies and the missing mass. *Astrophys. J.*, 303, 523-34.
[5] Tyson, J. A., *et al.* (June, 1984). Galaxy mass distribution from gravitational light deflection. *Astrophys. J.*, 281, L59-62.
[6] Lerner, E. J. (December, 1986). Magnetic Vortex Filaments, Universal Invariants and the Fundamental Constants. *IEEE Trans.*

110

Plasma Sci., Special Issue on Cosmic Plasma, PS-14, 6, 690-702; https://doi.org/10.1109/TPS.1986.4316620.

Using this value, we obtain a *predicted* T of 2.77 ± 0.13 K (assuming an uncertainty of ± 20% in n_1) compared with an *observed value* of 2.78 ± 0.03 K, (Johnson & Wilkinson 1987), an *excellent agreement*.

Current rates of *thermonuclear burning* appear to be inadequate to produce this much *energy* and *helium* in the *time since the origin of the galaxies* (about 17 Gy). [???] With a *mass/light ratio* of 20 M_s/L_s in the *visible, current stars appear to fall short by a factor of 33. Infrared production is comparable*, but the *combined brightness* still is short by a factor of nearly 20. The following section investigates *why earlier generations of stars were more massive and luminous and therefore produced far greater nuclear power.*

3. Gravitational Condensation by Plasma Vortices and the Formation of Massive Early Stars

A universal relationship exists between the distance, r, between condensed objects and the density, n, of the plasma from which they condensed. This relationship holds over 15 orders of magnitude of mass, from *stars* to *clusters of galaxies. The underlying reason for the existence of this relationship is the role played in the process of gravitational condensation by plasma vortices which have typical ion velocities* $(m_e/m^p)^{3/4}$ c. (Lerner, 1986[6]) Condensations are separated by distances r equal to the *ion collision distance* = 1.0 x 10^{19} n^{-1} cm, or nr = 1.0 x 10^{19} /cm². We can restate this relationship as M = 1.8 n^{-2} where M is the *mass* of the *condensed object* in *solar masses*. We thus see that *in the early stages of galactic condensation, when the average plasma density in the galaxy was lower than at present, the average mass of stars formed was higher.*

4. The abundance of helium and heavy elements

We can use this relationship in a simplified model of star formation to determine approximately how much helium and heavy elements were created during the formation of a galaxy. We begin with a cloud (considered cylindrical to simplify calculations) contracting in the axial direction, with *spiral radial magnetic filaments* in the *plane of rotation*, as hypothesized in earlier models (Lerner, 1986[6]; Peratt, 1986[7,8]).

[7] Peratt, A. L. (December, 1986). Evolution of the plasma universe. I. Double radio galaxies, quasars, and extragalactic jets. *IEEE Trans. Plasma Sci.*, PS-14, 639-60; https://doi.org/10.1109/TPS.1986.4316615.

[8]Peratt, A. L. (December, 1986). Evolution of the plasma universe. II. The formation of systems of galaxies. *IEEE Trans. Plasma Sci.*, PS-14, 763; https://doi.org/10.1109/TPS.1986.4316625.

The *filaments* carry *currents* in dense ropes toward the center and out along the axis. *As the main body of the plasma streams past the vortex filaments, shock waves are set up, rapidly compressing the plasma and initiating star formation.*

At any given instant, B/n is a constant throughout the *contracting plasma*. Since *the currents converge toward the center and flow out along the axis*, we have $B = 0.2 \, l/r$, where l is the *galactic current* and r is the *distance from the axis*. *Density*, n, thus also decreases outwards from the center, *with the heaviest stars forming in the least dense regions, furthest out.* We assume that the *incoming filaments* are sufficiently numerous (of the order of a dozen) that in any given *annulus* star formation is occurring *whenever previous generations of stars have released their gas to the interstellar medium* or have *not yet been formed.* That is, stars of an appropriate *mass* constitute essentially all the *plasma mass* in each *annulus*, so we can consider the situation simply along a single radial slice.

112

4.1. Process of stellar formation

For stars more massive than 12 Ms, which end their lives as *type I supernovae*, the *shock wave* spreads outwards vertically from the plane of rotation until broken up by *supernovae* explosions. The height of such a *shock wave* is thus just $L_t V_a$ where L_t is the *stellar lifeline* and V_a is the *velocity* of the *shock wave*, the *Alfven velocity* is the *plasma* at that point. *For less massive stars, which end thermonuclear burning less violently*, not becoming *supernovae*, the *shock wave* continues outwards throughout the entire cloud thickness. *This naturally biases element creation towards helium rather than heavier elements*, since, not coincidentally, *only the more massive stars generate significant amounts of carbon, oxygen and heavier elements.* (Arnett, 1978[9], Maeder, 1983[10], Adouze & Tinsley, 1976[11]).

[9] Arnett, W. D. (February, 1978). On the bulk yields of nucleosynthesis from massive stars. *Astrophys. J.*, 219, 1008-16; https://articles.adsabs.harvard.edu/pdf/1978ApJ...219.1008A.

[10] Maeder, A. (1983). *Proc. ESO Workshop on Primordial Helium*, 89.

[11] Adouze, J. & Tinsley, B. M. (1976). Chemical evolution of galaxies. *Ann. Rev. Ast. & Astrophys.*, 14, 43-79.

As the broader *shock waves* of the lighter M < 12 M, *stars* propagate outwards, the available *plasma* is tied up first by M = 12 M, *stars*. *Further star formation is only possible when these release the bulk of their mass back to the medium after a lifetime of about 25 My.* (Ezer & Cameron, 1971[12]).

[12] Ezer, D. & Cameron, A. G. W. (December, 1971). The evolution of hydrogen-helium stars. *Astrophys. Space Sci.*, 14, 399-421; https://doi.org/10.1007/BF00653327.

The stars with mass appropriate to the density at that point then monopolize the process again, thus setting up a small number of outward burning bands, each dominated by *stars* of a narrow range of *masses*.

Assuming that the *contraction of the cloud* is described by $H = e^{-t/t1} R_m$, where H is the *height of the cloud* along its axis, and R_m is its radius we have $n = n_i{}^{t/t1}$.

We can calculate the *mass* present as *stars of a given mass* M (in units of M_s)

$$\mathcal{M} = \int_{r1}{}^{r2} 4\pi R_m\, e^{t/t1}\, nm_p r\, dr \qquad (3)$$

using the relation $2nr = n_i R_m\, e^{-t/t1}$, derived from the fact that B/n is constant we can rewrite (3) as

$$\mathcal{M}/M_s = \tfrac{1}{2} \int_{n1}{}^{n2} n_i\, e^{t/t1} n^{-2}\, dn = \tfrac{1}{2} \int_{n1}{}^{n2} n_i\, e^{t/t1} n^{-1} \qquad (4)$$

where M, is the *total mass* of the *contracting cloud*.

We have $n_2 = 1.33/M^{0.5}$ and $n_1 = n_2 e^{-Lt/t1}$, the *density when stars of mass M form* and when they *expire* respectively. So

$$\mathcal{M}/M_s = 0.375\, m_i\, e^{t/t1}\, (1 - e^{-Lt/t1})\, M^{0.5}. \qquad (5)$$

4.2. Helium abundance

From *stellar evolution theory* we know the amount of *helium* produced for *stars* of various *masses,* varying from about 10% for $M = 12\ M_s$ to 3% for $M = 5\ M_s$. (Adouze & Tinsley, 1976[11])

$$He(M) = \int_{t0}{}^{tf} F(M) M^{0.5}\, n_i\, 0.375\, t_1 L_t^{-1} (1 - e^{-Lt/t1})\, e^{t/t1} \qquad (6)$$

where F is the *fraction of mass converted to helium by stars of mass* M and He(M) is the *total fraction of the galactic mass converted to helium by these stars*.

The *end time,* t_f, is just the *time at which the band of stars of mass M reaches radius* R_m. So $t_f = t_1(\ln 2.66/n_i\, M^{0.5})$. The *beginning time,* t_0, is defined by *the point at which the shock wave first forms*, that is *when V, the velocity of the plasma past the filament exceeds* V_a, the *Alfven velocity*.

We take V to be constant and equal to the orbital velocity of the galaxy, as is consistent with both observation and simulations (Peratt, 1986[7,8]) showing the flat rotation curves and as would be expected from a cylindrical object having nr constant.

$$V = (2\pi)^{0.5} R_m n_i^{0.5} G^{0.5} m_p^{0.5} \qquad (7)$$

Now, $V_a = I/10h^{0.5}n^{0.5}m_p^{0.5}r$ where m, is the *proton mass*, and I the *galactic current*. Studies by Beck, (1986)[13] have shown that galactic magnetic fields and currents can be related to a galaxy's mass per unit surface area, and thus to its orbital velocity.

[13] Beck, R. (December, 1986). Interstellar Magnetic Fields. *IEEE Trans. Plasma Sci.*, PS-14, 740-47; https://doi.org/10.1109/TPS.1986.4316623.

From Beck's results, we can derive the empirical relation

$$I = 1.5 \times 10^{-4} V^2 G^{-1}.$$

From this and the previously derived relations, and (7), we get:

$$V_a/V = 0.19 \, e^{-t/t1} M^{-0.25} n_i^{-0.5}. \qquad (8)$$

Thus $t_0 = \ln (0.19/M^{0.25} \, n_i^{0.5})$. Substituting into (6), we get

$$He(M) = F(M) \, t_1 L_t^{-1} (1 - e^{-Lt/t1}) \, (1 - 0.073 M^{0.25} n_i^{0.5}) \qquad (9)$$

What is t_1, the *characteristic time* of the *contraction*? As we shall show, *the contracting galaxy rapidly become sufficiently dusty to be opaque to the ultraviolet radiation emitted by these hot stars. Thus, the radiation energy density* must be balanced, as the *contraction* proceeds, by the *gravitational energy*. This can be expressed as

$$HeR_m/t_1c = \tfrac{1}{2} m_p V^2/E_h \qquad (10)$$

or *the thermal energy per particle equals the gravitational binding energy.*

For the Milky Way, we find that $V^2/R_m = 1.2 \times 10^{-8}$ cm/sec^2 and $n_i = 0.35$ (taking $M_g = 2 \times 10^{11}$ M_s, $R_m = 5 \times 10^{22}$ cm). Substituting these values into (9) and (10) integrating over a *mass* range $4M_s < M < 12$ M_s and solving the equations simultaneously, we get as a solution $t_1 = 8 \times 10^{15}$ second (260 My) and He = 0.225, *in excellent agreement with observation*. Given the uncertainty in the *mass* of the Galaxy a range of solutions is possible: He varies from 0.213 to 0.248 as V^2/R_m goes from 1.15-1.25 $\times 10^{-8}$ cm/sec^2.

For galaxies in general, I have previously shown on theoretical grounds that V^2/R *should be constant for early galaxies* (Lerner, 1986a[14]), and in fact actual *galaxies* have a relatively minor spread, which has undoubtedly been broadened in evolution by *tidal forces* and other later phenomena.

[14] Lerner, E. J. (May, 1986). Magnetic self-compression in laboratory plasmas, quasars and radio galaxies. Part I. *Laser and Particle Beams*, 4, 193-213; https://doi.org/10.1017/S0263034600001750.

These *tidal forces* can only act to *expand* the *galaxies* and thus decrease V^2/R_m or M_g/R_m^2 (which is proportional). So, we can estimate the *initial* V^2/R_m by looking at the maximum values for current galaxies. We find that over a broad range of *types* and *masses* these values fall between 1.0 and 1.25 $\times 10^{-8}$ cm/sec^2 (Lerner, 1986[6]). For a *mass* range from 2×10^{10} to 5×10^{13} M, we determine that He will vary over a range of 0.21 to 0.25, *in good agreement with observation*.

Most significantly, since He varies inversely with V^2/R_m, the observed upper limits on V^2/R_m sets lower limits to He and it is these minimum He values which have given reasonably consistent figures for 'primordial' helium abundance.

4.3. Carbon abundance

For $M > 12$ M_s, we must modify (4) by changing the height term from $R_m e^{-t/t1}$ to $V_a L_t$. We then get

$$(\mathrm{d}\mathcal{M}/\mathrm{dM})\ \mathrm{M_g}^{-1} = 0.125\ \mathrm{n_i}\ e^{t/t1}\ \mathrm{M}^{-1.75}. \tag{11}$$

The rate of production of carbon per unit mass for the *stellar mass* range $25\ \mathrm{M_s} < \mathrm{M} < 100\ \mathrm{M_s}$ is about $0.03/\mathrm{L_t}$ (Adouze & Tinsley, 1976[11]) and for this same range $\mathrm{L_t} = 9 \times 10^{15}/\mathrm{M}$.

So $\mathrm{C} = 3.2 \times 10^{-19}\ \mathrm{M(d}\mathcal{M}/\mathrm{dM)}\ \mathrm{M_g}^{-1}$ is the *fraction converted to carbon*. Integrating this over t and M we get

$$\mathrm{C} = (5.6 \times 10^{-19} - 1.1 \times 10^{-19}\ \mathrm{n_i}^{0.5})\mathrm{t_1}$$
$$= 4.5 \times 10^{-3} - 8.5 \times 10^{-4}\ \mathrm{n_i}^{0.5} = 0.0042 \tag{12}$$

This is in good agreement with observations of $0.004 - 0.005$.

4.4 Oxygen abundance

Using the same calculation for *oxygen*, and adjusting the rates of production accordingly we get a value of 0.018 which is somewhat high compared with observed values of 0.012 but is within the uncertainties generated by *stellar evolution theory*, especially for pure *hydrogen stars*, which we are here assuming. As He varies, we find that dHe/dZ = 3.3, again in good agreement with observation (Lequeux, *et al.*, 1979[15]).

[15] Lequeux, J., *et al.* (December, 1979). Chemical Composition and Evolution of Irregular and Blue Compact Galaxies. *Astronomy & Astrophysics*, 80, P, 155, 1979; https://adsabs.harvard.edu/full/ 1979A%26A....80..155L.

So, it is clear that *heavy elements* [*carbon* and *oxygen*] are not in fact overproduced by early *stars* compared with current observation.

We should note that this model does not yield for any parameters the very low heavy element abundances that are observed in *dwarf galaxies* like the SMC, combined with He of $0.22 - 0.24$ (Lequeux, *et al.*, 1979[15]). However, this does not contradict the model, since it is clear that He ions will be able to migrate out of the galaxies to enrich the immediate surrounding medium much more easily than the *heavier elements* such as *carbon* and *oxygen*. It is out of this He-enriched but *heavy element* poor medium that *dwarf galaxies* form. Similarly, He produced by the dwarfs above the local average level diffuses outwards, while only small amounts of *heavy elements* have yet had time to form.

4.5. Deuterium abundance

This same model can account for the observed amounts of *deuterium* in the galactic environment. Recent observations by Davis and others (Davis, 1986[16]) of substantial neutrino fluxes from solar flares indicate that the production of deuterium from p + p → d + π is much greater than theoretical calculations predict.

[16] Davis, R. (1986). Report on status of solar neutrino experiments. *Seventh Workshop on Grand Unification/Icoban '86*; https://www.gbv.de/dms/tib-ub-hannover/02249829X.pdf.

In fact, large flares with energy in *cosmic rays* of about 2 x 10^{34} ergs generate as many as 10^{38} *neutrinos*. Since each event releases the expenditure of at least 140 MeV in the production of a *pion*, more than half the total energy of the flare is going into *pions* (and *deuterium* production).

We can estimate the proportion of total *thermonuclear* release from *massive stars* that goes into *cosmic ray* production. Helou, *et al.* (1985)[17] point out that there is a close linear correlation between radio power generated by *galaxies* and *IR thermal radiation*, presumably derived from *young massive stars*.

[17] Helou, G., *et al.* (November, 1985). Thermal infrared and nonthermal radio: remarkable correlation in disks of galaxies. *Astrophys. J.*, 298, L7-11; https://articles.adsabs.harvard.edu/pdf/ 1985ApJ...298L...7H.

If we take as a measure of total *cosmic ray* production (twice radiated power) 3 x 10^{19} f (1.49 GHz) (the flux at that frequency) we find that about 1.2% of *thermonuclear* yield is in the form of *cosmic rays*. This yields 20 Kev per *hydrogen* atom in *cosmic ray energy*. If about 1 Gev of energy is used for each *deuterium* production, ½ of the energy goes into the production of *deuterium* and the current abundance should be in the area of 2 x 10^{-5}. The figure is thus *in good agreement with estimates of deuterium abundance* in the area of 2-3 x 10^{-5} (Steigman, 1985[18]).

[18] Steigman, G. (1985). Primordial nucleosynthesis — A window on the early universe. *Nuclear Physics*, B, 252, 11-24; https://doi.org/10.1016/0550-3213(85)90422-5.

5. The radiation of hot early stars

The radiation produced by these *hot early stars* is mostly in the *ultraviolet*. But such radiation is rapidly degraded to *infrared blackbody radiation* by *absorption* and *scattering* by particles in the *dusty plasmas* that make up *early galaxies*. In *current spirals*, *dust* is concentrated in a thin plane, but in the *embryonic galaxies* the *dust* is well mixed, obscuring *hot stars* in all directions.

With most of the radiation emitted by *stars* of around 10 *solar masses*, with *surface temperatures* of 40,000 K, we find that a *galaxy* like ours will become opaque when the *carbon/hydrogen mass ratio* is about a half percent of the present level. For this case, opacity occurs rapidly.

There is substantial observational evidence that embryonic galaxies do, in fact, go through this *early, bright massive-star* period, generating vast amounts of *IR radiation*. Allen & Roche, (1985)[19], using IRAS data, have found several dozen *galaxies* with *IR luminosities* between 10^{12} and 3 x 10^{13} that of the *sun*.

[19] Allen, D. A., Roche, P. F., & Norris, R. P. (March, 1985) Faint IRAS galaxies - A new species in the extragalactic zoo. *MNRAS*, 213, 67-74; https://www.adsabs.harvard.edu/full/1985MNRAS.213P..67A.

These galaxies appear to be of normal size with *luminosity/mass ratio* about 100 times that of the *sun*.

Our model predicts an *average luminosity/mass ratio* that is almost exactly the same—around 110 for t_1 of 8×10^{15}.

6. Isotropization of the microwave background

How is the initial infrared radiation scattered and isotropized throughout space to become the smooth microwave background? The basic process proposed here is *synchrotron absorption* and *remission* by Gev *electrons* trapped in the beams of *jets* emitted by *active galactic nuclei.*

[*Synchrotron radiation* (also known as *magnetobremsstrahlung*) is the *electromagnetic radiation* emitted when *relativistic charged particles* are subject to an acceleration perpendicular to their velocity (a ⊥ v). It is produced artificially in some types of particle accelerators or naturally by fast *electrons* moving through *magnetic fields*. The radiation produced in this way has a characteristic *polarization*, and the *frequencies* generated can range over a large portion of the electromagnetic spectrum.

Synchrotron radiation is similar to *bremsstrahlung radiation*, which is emitted by a *charged particle* when the acceleration is parallel to the direction of motion. The general term for radiation emitted by particles in a *magnetic field* is *gyromagnetic radiation*, for which *synchrotron radiation* is the *ultra-relativistic* special case. Radiation emitted by charged particles moving *non-relativistically* in a *magnetic field* is called *cyclotron emission*. For particles in the *mildly relativistic range* (≈ 85% of the speed of light), the *emission* is termed *gyro-synchrotron radiation*.

In *astrophysics*, *synchrotron emission* occurs, for instance, due to *ultra-relativistic motion* of a *charged particle* around a *black hole*. When the source follows a circular geodesic around the *black hole*, the *synchrotron radiation* occurs for orbits close to the photosphere where the motion is in the *ultra-relativistic regime*.]

Consider first a *jet* that is in energetic equilibrium with the *microwave background absorbing* and *emitting* through *synchrotron processes* the exact same blackbody spectrum and the same power. (I shall later show how such *equilibrium jets* arise from the *non-equilibrium jets* we can directly observe.) By Kirchhoff's law, such a *jet* must have an *optical depth* of unity.

> [*Kirchhoff's current law*, also called *Kirchhoff's first law*, or Kirchhoff's junction rule, states that, for any node (junction) in an electrical circuit, the sum of currents flowing into that node is equal to the sum of currents flowing out of that node.
>
> *Kirchhoff's voltage law*, also called *Kirchhoff's second law*, or Kirchhoff's loop rule, states that the directed sum of the potential differences (voltages) around any closed loop is zero.
>
> *Optical depth* is the natural logarithm of the ratio of *incident* to *transmitted* radiant power through a material. Thus, *the larger the optical depth, the smaller the amount of transmitted radiant power through the material*.]

By equating the power per unit area falling upon the *jet* with the *emitted synchrotron power* per unit area we get.

$$(e^5/3m^4c^7)E_jB^2VR = (2\pi^5/15h^3c^2)(kT)^4 \qquad (13)$$

with m the *mass of the electron* (gm), E_j the *energy density of the electrons in the jet* (erg/cm^3), B the jet magnetic field, eV the *mean energy of the electrons* (electron-volts), R the *radius of the jet* (cm) and T the *temperature of the microwave background* (K).

In deriving (13), I assume, in accordance with observation, that *jets are force-free, minimum-energy filaments*, with *electron velocity along magnetic field lines*. Electrons here have an average *effective energy*

121

above *minimum energy* of kT, where T is the *temperature of the radiation field*. (In contrast, for *thermalized plasmas* with randomly moving *electrons*, the average *effective energy* is close to eV, thus yielding much lower net *absorbtion*, since stimulated *emission* is higher.)

Assuming the *electron energy density* and the *magnetic energy density* to be equal,

$$E_j = B^2/8\pi \tag{14}$$

we can rewrite (13) as

$$(\gamma E_j^2/E_e)(R/r_e) = E_\lambda/8 \tag{15}$$

where γ is eV/mc^2, E_e is the *energy density* of the *classical electron*, or $(3/4\pi)(mc^2)^4/e^6$, r_e is the *classical radius of the electron*, and E_λ is the *energy density of the microwave background*. Or,

$$B^4VR = 7.7 \times 10^{10}T^4 \tag{16}$$

We also must have the *peak synchrotron frequency* equal to the *peak blackbody frequency*, or

$$eB\gamma^2/2\pi mc = AkT/h \tag{17}$$

where A is a numerical constant, about 6.7. This can be transformed into

$$\gamma^8 E_j^2/E_e = ..A^4\alpha E\lambda/8 \tag{18}$$

where α is the fine structure constant.

Solving (15) and (18) simultaneously we get

$$\gamma^7 = (5/2\pi^3)A^4\alpha R/r, \tag{19}$$

Since γ is not directly observable, we use (15), (18) and (19) to get R as a function of the *total emitted radiation from the galactic nucleus*, which we take to be equal to the total power of the *jet*, P

$$P = 2\pi R^2 E_j c. \tag{20}$$

We then have

$$R = 5^{0.05} 2^{0.65} A^{0.2} \alpha^{0.05} P^{0.7} / \pi^{0.5} c^{0.7} r_e^{0.4} E_\lambda^{0.35} = 7.0 \times 10^{-9} T^{-1.4} P^{0.7}. \tag{21}$$

6.1. Scattering of microwave photons

Since the *jets* have an *optical depth* of unity, a fraction $1 - 1/e$ of all background *photons* falling on the *jet* will be *absorbed*, while an equal number, of the same spectral distribution, will be *emitted*. We can consider this as a '*scattering*' of the *photons*, with complete randomization of direction in each such scattering.

The *scattering area* of the *jet* is just 2RcH, where H is the *time since the formation of galaxies and their active nuclei*. (It does not matter if an individual *nucleus* has been continuously active for this time as long as the *density* of such *nuclei* does not change.) The average *scattering length* of a *microwave photon* is

$$S^{-1} = 1.5 \times 10^{-8} \, cHT^{-1.4} \int D(P) \, P^{0.7} \, dP \tag{22}$$

where D(P) is the *density function of active nuclei* of *radiated power* P. D(P) can be derived from known data (Fanti & Perola, 1977[20]).

[20] Fanti, R. & Perola, G. C. (August, 1977). Luminosity Functions for Extragalactic Radio Sources. *Radio Astronomy and Cosmology; Proceedings of the Symposium, Cambridge University, Cambridge, England, August 16-20, 1976.* Edited by D. L. Jauncey. Symposium sponsored by the International Astronomical Union Dordrecht, D. Reidel Publishing Co. (IAU Symposium, No. 74), 171.

In doing so, we assume that H_0 = 60 Km/sec Mpc and that *total power* P(erg/sec) = 3 x 10^{19} P'(W/Hz at 408 MHz). We find we can closely approximate the observed data by

$$D(P') = 3 \times 10^{-65}\, P'^{-1.5}/cc, \qquad 10^{19} < P' < 10^{21} \qquad (23)$$
$$6 \times 10^{-66}\, P'^{-1.5}/cc, \qquad 10^{21} < P' < 10^{22}.$$

For P' > 10^{22}, the fall-off in *density* is sufficiently rapid to ignore the more powerful, but far less numerous, sources.

Taking H as 5 x 10^{17} sec (17 Gy), and assuming that the *galactic density* within *superclusters* is about 25 times the *average density* (Einasto, *et al.*, 1980[21]), we get S = 1 Mpc.

[21] Einasto, J., *et al.* (November, 1980). Structure of superclusters and supercluster formation. *MNRAS*, 193, 353-75; https://www.adsabs. harvard.edu/full/1980MNRAS.193..353E.

Thus, *if our hypothesis is valid, the microwave photons we see were last scattered a few million years ago, not 15 billion years ago.*

If we model the *local universe* as a collection of *supercluster filaments* each 6 Mpc in *radius* and with an average spacing of 50 Mpc (Einasto, *et al.*, 1980[19]), then a *photon* will encounter a *supercluster* about once in 300 My and will be scattered 200 times before again departing. Thus in *15 billion years* [???], a *photon* will travel only a few hundred Mpc from its point of origin and will be scattered more than 10^3 times, thoroughly isotropizing the radiation.

Initial anisotropy will be reduced by a factor of $e^{1,000}$. Additional *anisotropy* will be introduced by *inhomogenities* in the distribution of the *filaments* themselves. The most important *inhomogeneity* is that created by the approximately *cylindrical concentrations of mass and filaments in hyperclusters* about 200 Mpc in *radius* (Tully, 1986[3]). Observations in the direction perpendicular to the axis of such *concentrations* will include more of the dark areas between *filaments* than *observations* along the axis, which penetrate a greater depth of *filaments.* A fraction $\sim e^{-8}$ or 3×10^{-4} of the sky will be dark background between the *emitting filaments*. Since T varies as the ¼ power of *flux density*, we expect from this an average *anisotropy* of the *microwave background* at large angles of about 4×10^{-5}, *in good agreement with observation*, especially considering the rather large uncertainty in the *total optical depth* of the *hypercluster*.

It should be noted that the bulk motion of the *plasma* in the *jets* will be governed by *the powerful magnetic fields* in the *jets*, typically 10^{-4} G, *not primarily by the local gravitational fields*, so that large relative velocities between the *local microwave radiation field* and individual *galaxies*, such as our own, are to be expected.

We can calculate from this *scattering length* and geometry that the leakage rate from *local hyperclusters* would result in only a 40% decrease in *energy density* over a 15-17 billion-year period [???], even if the *hyperclusters* were assumed to be surrounded by empty space. Actual net leakage is clearly less, and has been made up for by additionally energy release, accounting for the current He abundance around 25% higher than the primordial one we used to calculate the *microwave temperature*.

6.2. Jets emitted from galactic nuclei

I have hypothesized the existence of such jets at equilibrium with the microwave background. How would such equilibrium be achieved? Initially, *jets* emanating from *galactic nuclei* will have *radii* far smaller than the 10 pc typical of *equilibrium filaments*, with far higher *magnetic fields* and *total radiation densities*. For example, *filaments* near our own *galaxies' nucleus*, which is not in an active state, have *radii* of about 0.5 pc and *magnetic fields* in excess of 10 mG (Yusef-Zadeh, *et al.*, 1984[22]; Aitken, *et al.*, 1986[23]).

[22] Yusef-Zadeh, (August, 1984). Large, highly organized radio structures near the galactic center. *Nature*, 310, 557–61; https://doi.org/10.1038/310557a0.

[23] Aitken, D. K., *et al.* (January, 1986). Infrared spectropolarimetry of the Galactic Centre: magnetic alignment in the discrete sources. *MNRAS*, 218, 363-84. https://doi.org/10.1093/MNRAS/218.2.363.

Initially radiation from *electrons* in these *jets*, accelerated locally by *plasma* process (such as interaction with *proton beams*), (Lerner, 1986a[14]), radiate far more *energy* than they absorb from the *microwave background*. However, as the *jet* expands, the *radiation density* falls until it equals that of the background. At this point the *electrons* at appropriate energies to absorb from the background (typically around 25 Gev) are in equilibrium. *Radiation pressure* is also balanced, ending the radial expansion of the *jet*. More energetic *electrons* loose energy on net, radiating faster, while less energetic ones gain energy on net, thus maintaining an equilibrium energy distribution centered on the *resonant electron energy*. Eventually all the *energy* of the beam is trapped in this *electron distribution* which then carries the *current* of the *jet*.

The *jets*, once they reach equilibrium with the background are radio-invisible, since they radiate with the same spectrum as the background. *What evidence is there that they in fact exist?*

The most direct evidence is in the *observed relationship between the power of galactic nuclei and the observed length of their jets*. Our theory predicts that the *jets* will be observable only until their radii reach the equilibrium value defined by (21). Since it is known that such *jets* have a *length/radius ratio* of close to 10^2 (Banhatti, 1987[24]).

[24] Banhatti, D. G. (March, 1987). The width of jets in powerful edge-brightened extragalactic double radio sources. *MNRAS*, 225, 3, 487-90; https://doi.org/10.1093/mnras/225.3.487.

$$L = 7.0 \times 10^{-7} \, T^{-1.4} \, P^{0.7} \tag{24}$$

...

Figure 1. *Lengths of radio jets* (Kapahi, 1977) *show a clear correlation between maximum jet length and radio source power, in good agreement with our model prediction* (solid line), taking P = 2 x 10^{19} P' (178 MHz).

As shown in figure 1, *this relationship gives a close fit to the observed maximum observed lengths.* (*Jets* can, of course, have much shorter *observed jets* if the jet has only recently 'turned one', which is generally the case *since jets appear to be intermittent.*)

High resolution observations of *Cygnus A* (Perley, *et al.* 1984[25]) appear to show the transition between the highly organized, linear (non-equilibrium) *filaments* of the *jet* and the increasingly randomized and tangled *filaments* of the *lobe* which fade out in the background as they approach equilibrium.

[25] Perley, R. A., *et al.* (October, 1984). The jet and filaments in Cygnus A. *Astrophys. J.*, 285, L35-8; https://doi.org/10.1086/184360.

6.3. Inverse Compton radiated power and the gamma ray background

There is additional evidence of the existence of the extended equilibrium *jets*. The concentrated Gev electrons are observable through their *inverse Compton collisions* with the *background radiation field*, producing a background *soft gamma-ray radiation*.

From (19), (21) and (23) we can calculate that the *electron energy distribution peaks* at

$$\gamma = 3.6 \ P^{01} \tag{25}$$

or around 22 Gev. The peak of the *energy spectrum* produced by the *inverse scattering* of these *electrons* from the 2.75 K background is proportional to $\gamma^2 T$. If we express the *spectrum* as *energy* per logarithmic interval, as in figure 2, the peak of the inverse Compton spectra is 2.2 x 10^{-9} γ^2 Mev or 4.2 Mev (this is 1.6 Mev peak for a conventional w/Hz spectrum).

...

Figure 2. *X-ray and gamma-ray extragalactic background* radiation is plotted in terms of power per unit logarithmic interval. Data from Mazets, 1975(x) and Schonfelder, 1983(.). Hump in Mev range, rising above extrapolation from x-ray background *is in excellent agreement with model predictions* (solid curve).

The total power generated by *inverse Compton radiation* is

$$P_i = (1/2)\pi^2 c^3 HE^2 R_{sc} \int R^3 P^{-1} D(P) \, dP \text{ erg/sec cm}^2 \quad (26)$$

where R_{sc} is the *radius of the supercluster*, 6 Mpc. This is about 0.1 Mev/cm^{2}sec-sr, including contributions from nearby *superclusters*. Since there is good evidence that *the large-scale density of matter is concentrated locally*, within around 200 Mpc of the Milky Way, to a level at least four times the universal mean, (Tully, 1986[3]) we find that the *inverse Compton power* contributed by *filaments* outside this *concentration* is relatively small. From our hypothesis that *the microwave background radiation is produced by thermonuclear processes in early galaxies*, the *local microwave energy density* is proportional to the *local matter density*. The efficient *scattering* in the denser regions will prevent the radiation from equally distributing itself over distances larger than a few hundred Mpc. From (26) we can conclude that the *power density* of the *inverse Compton radiation* varies as n^2. Thus, the contribution from most of the universe, where n is less than 0:25n (*local*) will be small compared to the contribution from nearby *supercluster filaments*.

A comparison of calculated and observed gamma ray background (figure 2) *shows excellent agreement*. This is particularly significant since *no Big Bang explanation has been successfully proposed for the well-known low-Mev 'bump'*, which is here naturally accounted for.

7. Absorption by filaments and the spectra of radio sources

The *filaments* also reveal themselves as effective *absorbers* of radio waves from distant sources. Their *opacity* must increase rapidly beyond about 2mm *wavelength*, with *transmitted power* dropping 10-fold by 5mm and another 20-fold by 1cm. By 2cm, the *filaments* are essentially *totally opaque*, having a *transmittance* of less than 10^{-9}. Fortunately for radio astronomers, the *filaments* do not cover the sky thoroughly. Looking outwards from a *supercluster filament*, about e^{-3} of the sky, or 5%, would not be obscured by a *filament*.

7.1. Obscuration effect on compact and extended sources

Obscuration by *filaments* will affect *compact* and *extended radio sources* differently. *Compact sources* (unresolved by the *Very Large Array* (VLA) will either by covered by a *filament* or shine through undimmed between the *filaments*—thus its radio emission will be either entirely suppressed or unaffected. *Extended sources* will, in contrast, be uniformly suppressed by a factor of 20. (The fineness of the *filaments* will make it unlikely, although not impossible, that they would be resolved directly crossing an extended source.)
…

Compact radio sources will have their numbers suppressed, since those behind *filament* will be totally radio quiet. Only 5% of nearby *optical sources* can be expected to have *compact (flat spectrum) radio emission*, in good agreement with observation (Peacock, 1985[26]).

[26] Peacock, J. A. (December, 1985). The high-redshift evolution of radio galaxies and quasars. *MNRAS*, 217, 601-31; https://doi.org/10.1093/mnras/217.3.601.

For more distant ones, only those not obscured by intervening *supercluster filaments* will be visible and only the brightest, from the center of their own *cluster*, will be observable in any case, so the *apparent abundance*, relative to the *optical sources* should fall to 3×10^{-4} for distant sources, *again in agreement with observation*. Thus, the puzzling statistics of *quasars, flat spectrum* and *steep spectrum sources*, are neatly explained.

7.2. Optical polarization

Finally, we note that *optical polarization* will also be suppressed for objects behind *supercluster filaments*. The *magnetic field* created by the *jet filaments* collectively, amounting to an average of about 0.2 microgauss in a *supercluster*, will align *dust particles* with the *net field* if

$$B^2/n > 0.26 \, aT_d(3_kT_g)^{1/2}$$

where a is the *dust particle diameter*, T_d the *dust temperature*, T_g the *gas or plasma temperature* (Aitken, 1986[21]). This condition will hold assuming $T_d = 3K$, $T_g = 10$ keV and $a < 2.5 \times 10^{-5}$ cm, all being reasonable assumptions.

The *supercluster* material obscuring the radio-quiet sources, amounting to about $2 - 4 \times 10^{20}/cm^2$, contain enough *dust* to absorb roughly 25% of the light from distant sources (Phillips, 1986[27]).

[27] Phillips, S. (1986). *Astrophys. J. Lett.*, 25, 19.

Thus, for reasonable levels of *polarization* of the *grains, intrinsic polarization* of up to 10-12% could be wiped out. This corresponds to the observation that *radio-loud* sources have optical polarizations of 2-15% but *radio-quiet* sources are generally less than 1% *polarized* (Landau, *et al.* 1986[28]).

[28] Landau, R., *et al.* (September, 1986). Active Extragalactic Sources: Nearly Simultaneous Observations from 20 Centimeters to 1400 Angstrom. *Astrophys. J.*, 308, 78; https://doi.org/10.1086/164480.

8. Conclusions

The *microwave background*, as well as the *abundance of helium and deuterium* can be explained naturally from *plasma* and *thermonuclear* phenomena *without recourse to the Big Bang. Given this, how does the Big Bang stand up as a testable scientific theory?* The most serious contradiction to the *Big Bang* is the now-confirmed existence of elongated *hyperclusters* in excess of 600 Mpc in length (Tully, 1986[3]; Shaver, 1987[29]).

[29] Shaver, P. A. (April, 1987) Possible large-scale structure at redshifts ≲ 0.5. *Nature*, 326, 773–5; https://doi.org/10.1038/326773a0.

Such objects must have taken at least a few hundred billion years to form, far longer than the time allowed in the Big Bang theory.

In addition, the observed large scale streaming velocities of galaxies (Collins, 1986[30]), the absence of 'dark matter' (Valtonen & Byrd, 1986[4]), the new and lower observed abundances of helium and deuterium (Ferland, 1986[31]; Vidal-Madjar & Gry, 1984[32]), and the low level of small-scale fluctuation in the microwave background all pose acute problems for the *Big Bang*.

[30] Collins, C. A. (April, 1986). Large-scale anisotropy in the Hubble flow. *Nature*, 320, 6062, 506-8; https://doi.org/10.1038/320506a010.1038/320506a0.

[31] Ferland, G. J. (November, 1986). Collisional Effects in the Helium Triplets, and the Primordial Helium Abundance. *Astrophys. J. Lett.*, 310, L67; https://doi.org/10.1086/184783.

[32] Vidal-Madjar & Gry, (August, 1984). Deuterium, helium, and the Big-Bang nucleosynthesis. *Astronomy & Astrophysics*, 138, 285-9; https://www.researchgate.net/publication/234303491_Deuterium_helium_and_the_Big-Bang_nucleosynthesis.

Other phenomena, such as the *gamma ray background*, and the previously cited *relation of density and mass* are unexplained and unpredicted by the *Big Bang*. In contrast, all of these phenomena can be predicted and quantitatively explained by *plasma hypotheses* based on the phenomena of *plasma filaments*.

At present, the underlying cause of the *Hubble expansion* remains uncertain. Yet, if cosmology is to be an effort to understand observations, not just an exercise in ideology, *the time has come for astrophysicists to abandon the Big Bang theory* and look elsewhere for explanations of the structure and evolution of the universe.

Lerner, E. J. (October, 1992). The Big Bang Never Happened—A Reassessment of the Galactic Origin of Light Elements (GOLE) Hypothesis and its Implications.

Book (569 pages): Knopf Doubleday Publishing Group; https://www.lppfusion.com/storage/GOLE-Lerner.pdf.

Far-ranging and provocative, The *Big Bang Never Happened* is more than a critique of one of the primary theories of astronomy -- that the universe appeared out of nothingness in a single cataclysmic explosion ten to twenty billion years ago. Drawing on new discoveries in particle physics and thermodynamics as well as on readings in history and philosophy, Eric J. Lerner confronts the values behind the *Big Bang theory*: the belief that mathematical formulae are superior to empirical observation; that the universe is finite and decaying; and that it could only come into being through some outside force. With inspiring boldness and scientific rigor, he offers a brilliantly orchestrated argument that generates explosive intellectual debate.

Lerner, E. J. (1993). The Case Against the Big Bang.

In: Arp, H. C., Keys, C. R., Rudnicki, K. (eds) *Progress in New Cosmologies*. Springer, Boston, MA; https://doi.org/10.1007/978-1-4899-1225-1_7.

Abstract

Despite its widespread acceptance, the *Big Bang* theory *is presently without any observational support*. All of its quantitative predictions are contradicted by observation, and none are supported by the data. Its *predictions of light element abundances* are inconsistent with the latest data. It is impossible to produce a *Big Bang "age of the universe"* which is old enough to allow the development of the observed large-scale structures, or even the evolution of the Milky Way galaxy. The theory does not predict an isotropic *cosmic microwave background* without several additional ad hoc assumptions which are themselves clearly contradicted by observation. By contrast, *plasma cosmology* theories have provided *explanations of the light element abundances*, the *origin of large-scale structure* and the *cosmic microwave background* that accord with observation. It is time to abandon the *Big Bang* and seek other explanations of the *Hubble relationship*.

[*Plasma cosmology* is a non-standard cosmology whose central postulate is that *the dynamics of ionized gases and plasmas play important, if not dominant, roles in the physics of the universe at interstellar and intergalactic scales*[1].

[1] Alfvén, H. O. G. (1990). Cosmology in the plasma universe – an introductory exposition. *IEEE Transactions on Plasma Science*, 18, 5–10; https://doi.org/10.1109/27.45495.

In contrast, the current observations and models of cosmologists and astrophysicists explain the formation, development, and evolution of large-scale structures as dominated by *gravity* (including its formulation in Albert Einstein's *general theory of relativity*).

Developing ideas due to his senior colleague, the theoretical physicist Oskar Klein, Hannes Alfvén arrived in the early 1960s at a cosmological view which differed from both big-bang and steady-state cosmology. Klein and Alfvén rejected the big-bang theory, which they found unscientific and mythical. *Alfvén–Klein cosmology*, holds that *matter* and *antimatter* exist in equal quantities at very large scales, that the universe is eternal rather than bounded in time by the *Big Bang*, and that the *expansion of the observable universe* is caused by annihilation between *matter* and *antimatter* rather than a mechanism like *cosmic inflation*.

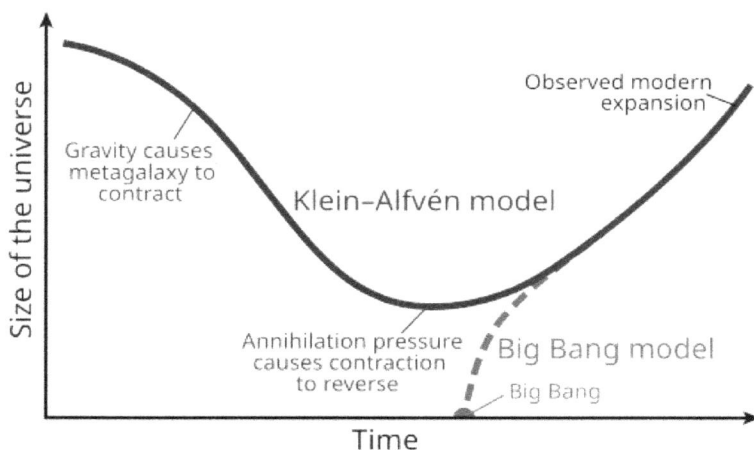

Cosmologists and astrophysicists who evaluated *plasma cosmology* and rejected it, claimed that it did not match the observations of astrophysical phenomena as well as the currently accepted *Big Bang* model.]

Ferrara, A., & Dettmar, R. -J. (May, 1994). Radio-emitting dust in the free electron layer of spiral galaxies: Testing the disk/halo interface.

Astrophys. J., 427, 155-9; https://doi.org/10.1086/174128; also at https://articles.adsabs.harvard.edu/pdf/1994ApJ...427..155F.

Received July 26, 1993.
Accepted November 22, 1993.

Abstract

We present a study of the *radio emission from rotating, charged dust grains immersed in the ionized gas constituting the thick, Hα-emitting disk of many spiral galaxies*. Using up-to-date optical constants, the *charge* on the *grains* exposed to the diffuse *galactic UV flux* has been calculated. An analytical approximation for the *grain charge* has been derived, which is then used to obtain the *grain rotation frequency*. *Grains are found to have substantial radio emission peaked at a cutoff frequency in the range 10-100~GHz*, depending on the *grain size distribution* and on the efficiency of the radiative damping of the *grain* rotation. The *dust* radio emission is compared to the *free-free emission* from the ionized gas component; some constraints on the magnetic field strength in the *observed dusty filaments* are also discussed. The model can be used to test the disk-halo interface environment in *spiral galaxies*, to determine the amount and size distribution of *dust* in their ionized component, and to investigate the rotation mechanisms for the *dust*. Numerical estimates are given for experimental purposes.

136

Lundin, R., & Marklund, G. (January, 1995). Plasma vortex structures and the evolution of the solar system— the legacy of Hannes Alfvén.

Physica Scripta, 60, 198-205; https://doi.org/10.1088/0031-8949/1995/ T60/023.

Abstract

In this report we address the present knowledge of *plasma vortex structures* and how that impacts on *cosmogony*, the *evolution of the solar system*, as it was conceived by Hannes Alfvén in 1954. Although *plasma vortex structures* are well known phenomena in contemporary *space plasma physics*, few attempts have been made to associate *plasma vortices* with large astrophysical objects. Indeed, Hannes Alfvén was a pioneer in this respect by associating the *early solar system* with a *plasma vortex*. Despite obvious merits from an observational and theoretical point of view, Alfvén-cosmogony based on a synthesis between *space plasma physics* and contemporary *planetology* have had few proponents. In fact, few people outside the *space plasma physics* community appear to appreciate the evolutionary thesis by Hannes Alfvén and his coworker Gustaf Arrhenius.

Alfvén cosmogony is based on three principles. (1) The evolution of the *solar nebula* into the *early sun* and its *planets* was a *plasma-physics process where electromagnetic forces combine with gravitational forces*. (2) *Chemical separation of matter* in the solar system was governed by the *critical velocity ionization effect*. (3) The *transfer of angular momentum* from the *magnetized sun* to the *planets* was due to *partial plasma corotation* in the *early solar nebula*. In this report some important aspects of *plasma vortex structures* and the *Alfvén cosmogony* will be addressed and it will be shown that a number of new observations within the *plasma* environment of *planets* and in *interplanetary space* corroborate *cosmogony* as envisaged by Hannes Alfvén.

Lerner, E. J. (May, 1995). Intergalactic Radio Absorption and the COBE Data.

Astrophys. Space Sci., 227, 1-2, 61–81; https://doi.org/10.1007/ BF00678067.

Abstract

The COBE data on *cosmic background radiation* (CBR) *isotropy* and *spectrum* are generally considered to be explicable only in the context of the *Big Bang theory* and to be confirmation of that theory. *However, this data can also be explained by an alternative, non-Big Bang model which hypothesizes an intergalactic radio-absorbing and scattering medium.* A simple, inhomogenous model of such an *absorbing medium* can reproduce both the *isotropy* and *spectrum* of the CBR within the limits observed by COBE, and in fact *gives a better to fit to the spectrum observations than does a pure blackbody.* Such a model does not contradict any other observations, such as the existence of distant radio sources.

---

1 Introduction

The Planck spectrum and high isotropy of the *cosmic background radiation* (CBR) has long been widely considered as the key observational support for the *Big Bang*. However, in 1988 the author proposed that the CBR was *isotropized* and *thermalized* by *synchrotron absorption* and *reemission* of *electrons* trapped in *dense magnetically pinched filaments* in the *intergalactic medium* (IGM)[1].

[1] Lerner, E. J. (August, 1988). Plasma model of microwave background and primordial elements: an alternative to the big bang. *Laser and Particle Beams*, 6, 456-69; https://doi.org/10.1017/ S0263034600005395.

A similar theory was independently put forward by Peter and Peratt (1990)[2].

[2] Peter, W., & Peratt, A. L. (1990). *IEEE Trans. Plasma Sci.*, 18, 49.

In the author's theory, the *energy* for the *cosmic background radiation* (CBR) was supplied not by the *Big Bang*, but by *hot, intermediate mass stars* in *early galaxies*, which also produced the observed *"primordial" helium*.

Observational evidence for the existence of such an *absorbing intergalactic medium* (IGM) came from studies of *radio* and *infrared brightness* of *galaxies* at various *distances* from Earth[3,4].

[3] Lerner, E. J. (September, 1990). Radio absorption by the intergalactic medium. *Astrophys. J.*, 361, 63-8; https://adsabs.harvard.edu/full/1990ApJ...361...63L.

[4] Lerner, E. J. (September, 1993). Confirmation of radio absorption by the intergalactic medium. *Astrophys. & Space Sci.*, 207, 17-26; https://doi.org/10.1007/BF00659126.

For a given *infrared brightness*, more distant *galaxies* were systematically dimmer in the *radio*, indicating a selective *absorption* of *radio frequency radiation*.

However, the new data provided by COBE (Bennett, *et al.*, 1994)[5] measuring the *anisotropy* of the *cosmic background radiation* (CBR) and the detailed spectrum (Mather, *et al.*, 1993)[6], present a challenge for any theory.

[5] Bennett, C. L., *et al.* (1994). *Astrophys. J.*, submitted.

[6] Mather, J. C., *et al.* (1993). Scientific results from the Cosmic Background Explorer (COBE). PNAS, 4766-73; https://doi.org/10.1073/Pnas.90.11.4766

In particular, with an *absorbing, emitting medium*, subject to the *Hubble relation*, radiation from more distant parts of the *medium* will be *redshifted*, so the sum of the radiation from the medium will not be a precise *blackbody*. Can a realistic model of this sort produce a spectrum as close to a *blackbody* as that observed by COBE?

In addition, parts of the *medium* that are *denser* than others, or slightly *hotter*, should produce *anisotropies*. Can these approximate those detected in the COBE data?

The present paper is intended to give a preliminary answer to these questions. However, it should be noted that the *Big Bang* itself does not provide a very satisfactory explanation of the Planckian spectrum and isotropy observed. As has been known for more than a decade, the original version of the theory does not predict even rough isotropy of the *cosmic background radiation* (CBR), since portions of the sky separated by more than about 10 degrees would not have had time to be in contact with each other and so could not have reached thermal equilibrium with each other. Neither isotropy nor a Planck spectrum would thus be expected on large angular scales, in gross contradiction to observation. (This is the so-called *horizon problem*). The widely accepted solution to this problem is *the additional hypothesis of an inflationary stage that expands a small portion of the universe that has reached thermal equilibrium.* This is a wholly ad hoc hypothesis, since it is derived from grand unified theories that are themselves without any experimental justification. Indeed, the only unequivocal predictions of these theories, the instability of the *proton*, has been contradicted by experiment[7].

[7] Becker-Szendy, R., et al. (November, 1990). Search for proton decay into $e^+ + \pi^0$ in the IMB-3 detector. *Phys. Rev.*, D, 42, 2974; https://doi.org/10.1103/PhysRevD.42.2974.

If inflation is accepted, it predicts an omega of unity for the *density* of the universe. Since the directly observed density leads to an omega of only 0.02 at most, *this is another violent contradiction with observation.* Indeed, an omega of unity with *ordinary matter* is inconsistent with *Big*

Bang nucleosynthesis predictions which allow at most a density of 0.04[8].

[8] Olive, K., *et al*. (March, 1990). Big-bang nucleosynthesis revisited. *Phys. Rev. Lett*, B, 236, 4, 454-60; https://doi.org/10.1016/0370-2693(90)90382-G.

To overcome this inconsistency and contradiction with observation, a third ad hoc hypothesis has been widely proposed that a new and wholly unobserved type of matter, *dark matter*, constitutes 98% of the universe. Again, the existence of such *dark matter* in the required amounts is utterly *without experimental or observational support*.

So, the *Big Bang* does not provide a good explanation of the COBE results, despite claims to the contrary. Does an *absorption model*?

2 Theoretical Model

To calculate the spectrum that results from the hypothesized *intergalactic medium* (IGM), three elements are required. *First*, we need an explicit, theoretically derived expression for the *absorptivity* of a given *density* of *filaments* for all *wavelengths* observed by COBE. *Second*, we need a hypothesized model of the *variation* of the *density* of the IGM with *distance*. *Third*, we need a formula to relate the *radiation received from a filament to the distance to the given filament*.

The *absorbent medium* has been described analytically previously (Lerner, 1992)[9].

[9] Lerner, E. J. (December, 1992). Force-Free Magnetic Filaments and the Cosmic Background Radiation. *IEEE Trans. Plasma Sci.*, 20, 6, 935-8; https://ieeexplore.ieee.org/document/199554.

The wavelength that the *filaments* are opaque to depends of the *radii* of the individual *filaments*. The ensemble of *filaments* acts as essentially a *perfect blackbody* in a wavelength range extending from about 2000 μ to 200 μ. They become slightly transparent by 800 μ and then increasingly transparent up to 200 μ, where they are completely transparent to radiation. *The model hypothesizes that the density of the filaments in space is approximately proportional to the energy density of the intergalactic magnetic field and to the smoothed density of matter.*

To summarize this model, we begin by hypothesizing *force free filaments* which are *electrically neutral,* and have *magnetic field energy* equal to *ion kinetic energy. Electrons* and *ions* travel along *field lines, absorbing* and *emitting synchrotron radiation.*

[*Synchrotron radiation* (also known as *magnetobremsstrahlung*) is the *electromagnetic radiation* emitted when *relativistic charged particles* are subject to an acceleration perpendicular to their velocity (a ⊥ v). It is produced artificially in some types of particle accelerators or naturally by fast *electrons* moving through *magnetic fields.* The radiation produced in this way has a characteristic *polarization,* and the *frequencies* generated can range over a large portion of the electromagnetic spectrum.]

In equilibrium with a *background radiation field, electrons* have a *temperature* T, perpendicular to the field lines. Following Lerner (1992)[9] we find that the *optical depth* A at the *synchrotron wavelength* A of each individual *filament* with *radius* R (cm) and *magnetic field* B (gauss) is

$$A = \lambda^2 J/2\pi k T_p = \lambda^2 n_e B^2 \gamma_e h r_e^2 R f /3mkT_p c \qquad (1)$$

where J is *emissivity*, n_e is *electron density*, r_e is the classical *electron radius*, m is the *electron mass* and λ is the *synchrotron wavelength* and

$$f = abs(v dR/R dv),$$

where $v = c /\lambda$.

[*Optical depth* is the natural logarithm of the ratio of *incident* to *transmitted* radiant power through a material. Thus, *the larger the optical depth, the smaller the amount of transmitted radiant power through the material.*]

This equation is simply *Kirchoff's law* relating *absorptivity* and *emissivity*, with J being the standard *synchrotron radiation emissivity*. (This formulation is valid when $kT_p < h\nu$, so that *electrons* are predominantly at the zero *energy-state* relative to the *magnetic lines of force*, and "jump" off them to the *energy* $h\nu$ when they *absorb* a *photon*. In the long wavelength limit $kT_p \gg h\nu$, the formula converges on the classical equation which is larger by a factor of $kT/h\nu$, since then the average *electron* spends most of its time near an *energy* of kT). For kT, $\ll mc^2$, which we'll see is the case for these *electrons* traveling along field lines, $v = eB\gamma_e/2\pi mc$ (γ_e is the *relativistic* factor). This is smaller than the normal formula $eB\gamma_e^2/2\pi mc$ for *synchrotron radiation* from *randomly oriented electrons* and applies to radiation *emitted in the forward direction*, as nearly all the radiation is.

Thus,

$$\lambda = 2\pi mc^2/eB\gamma_e. \tag{2}$$

With *magnetic field energy* equal to *ion energy*, $n_p = n_e = B^2/8\pi\gamma_p Mc^2$ where M is the *proton mass*. Substituting these values of n_e and λ into (1) we get

$$A = \pi h r_e B^2 R f/6kT_p Mc\gamma_e\gamma_p. \tag{3}$$

To find the values in (3), we must look at the formation of the *filaments*. As detailed in (Lerner, 1992)[9] these *filaments* are generated in *quasars*, *active galactic nuclei* and *Herbig Haro objects* when beams are accelerated by *high electric fields*. If a *proton* beam is to accelerate to energy $\gamma_p Mc^2$ in a *length* r, *energy loss* by *synchrotron radiation* will balance *energy gained* by *acceleration* if

$$r/c = 24Mc/r_p^2 B^2\gamma_p \tag{4}$$

or

$$I^3/r = 3(20)^3 e^3/2r_p^4 = 2.95 \times 10^{39} \text{ Amp}^3/\text{cm}. \tag{5}$$

Where I is *beam current* and r_p is the classical *proton radius*. This is equivalent to *beam power* $P_c = 2.8 \times 10^{34} r^{2/3}$ erg/sec. For *quasars* with r $= 10^{18}$ cm, $P_c = 2.8 \times 10^{46}$ erg/sec. For *active galactic nuclei* (AGN) with $r = 10^{15}$ cm, $P_c = 2.8 \times 10^{44}$ erg/sec.

…

The second part of the theoretical model is a *density function* which describes the *density* of *absorbers* as a function of *distance from the observer*. In the *general filament model*, since the *individual filaments* are essentially *perfect absorbers* and *scatterers*, the *absorbency* is a function of the portion of the sky covered by *filaments*, which increases with distance from Earth. A hierarchical bundle of *filaments* should, according to the model developed above have an *optical depth* of unity, viewed from the outside.

However, a realistic model must take into account the *observed inhomogenity* of the *universe*. *Observations of galactic distribution indicate that our galaxy is located inside an extensive filament of galaxies*, in relatively dense central portion of that *filament*. Thus, we would expect that the concentration of *absorbing filaments*, proportional to B^2, should decrease with distance from Earth, at least within this broad *supercluster of galaxies*. Indeed, observations of the *absorption* of *radio radiation* form other *galaxies* indicates that the *density* of *absorbers* declines as 1/D, with D the *distance form Earth* (Lerner, 1990, 1993)[3,4].

If we are located within a *filament*, or more precisely a roughly symmetrical concentration of *matter* and *fields* within such a *filament*, and if we assume that the *absorbing filaments* are randomly oriented within the much large *supercluster filament*, then we expect that total *optical depth* A' will be defined by

$$dA'/dD = 1/3D \text{ or } A' = \ln D/3. \tag{32}$$

This would lead to an expectation that *apparent radio luminosity* will fall as $D^{-1/3}$, in excellent agreement with observations of $D^{-0.32}$. Thus, one part of our *density model* defines the *density* of *absorbers* as $0.32/D$.
...

The third element of the model defines the *spectrum* and *intensity* of *radiation* from *absorbing filaments* at a given *distance* D. This depends on the actual mechanism that creates the *Hubble relation* in the first place. *In the Big Bang formulation* [???], the *spectrum* of a *blackbody radiation* at a *redshift* z is

$$J = 2\pi h v^3/c^2 [e^{\{hv(1+z)/kT\}} - 1]. \tag{33}$$

Here, the resulting spectrum remains a blackbody, and the total flux received drops as $(1 + z)^{-4}$. However, *other explanations of the Hubble relation produce different formulae.* For example, if the Hubble relation is created by a *true expansion in space* [???], then the *Doppler formula* is used which produce the relation

$$J = 2\pi h v^3 (1 + z)^2/c^2 [e^{\{hv(1+z)/kT\}} - 1] \tag{34}$$

Here the redshifted spectrum is no longer a *black body* and the total flux drops as $(1+z)^{-2}$.

[Tired-light theory?]

There are two physical processes expected from the model that must be taken into account. Both are a result of the *inhomogenity* the model assumes. A *concentration* of *filaments with a higher energy and matter density* than the surroundings *will expand*, driven by radiation pressure on the *filaments* and the surrounding *plasma*, and will cool with time. *This means that* (33) *must be modified.* If a body is *expanding uniformly*, a *Doppler shift* is introduced in addition to the *Hubble shift.*
...

3 COBE Data

Analysis of the COBE data by the initial researchers has yielded two conclusions: 1) that the spectrum *does not deviate from a blackbody* by more than 10^{-4} of peak brightness and 2) that *there are anisotropies* at a scale of 10 degrees of the order of 10^{-5}. For the purposes of this paper, we will accept the second conclusion as valid, even though it is still subject to debate. *The first conclusion, however warrants a closer look.*

The reduced data reported by Mather, *et al.* (1993)[6] does not fully support the conclusion that no deviations were detected. In fact, small deviations were detected, *outside the calculated experimental errors.* However, Mather, *et al.* rather arbitrarily increased these statistical errors by a factor of 2.3 to produce an undeviated fit to a *blackbody*. Since relatively small deviations in the spectrum will be important in testing the present theoretical model, a re-analysis of the data is presented here.

…

4 Results-Predicted Spectrum

To determine the prediction of the theoretical model for the *cosmic background radiation* (CBR) spectrum, we simply integrate over the hypothesized *density distribution, redshifting* the radiation.

…

This is an important result. *With no adjustable parameters other than the dimensions of the local matter concentrations, the observed COBE CBR spectrum is excellently fit by a model that assumes the cosmic background radiation* (CBR) *is produced by an intergalactic array of absorbing and emitting magnetic plasma filaments, which obscures distant radio sources by no more than a factor of several thousand, and fits all other observational constraints.*

At *frequencies* below the range studied by COBE, the model predictions remain well within the observational errors of other experiments. …

5 Results-Predicted Anisotropy

The *anisotropy* of the *cosmic background radiation* (CBR) predicted by this model will be affected by a number of phenomena: the *location of our galaxy relative to the center of the mass concentration* (the region within the cutoff C), *its motion, deviations of the concentration from sphericity and density and temperature variations with the concentration*. Since our *galaxy* will participate in the *expansion* of the *concentration* [???], its *velocity* relative to the center of motion of the *concentration* will be aHD, where D is the distance between our *galaxy* and the center of the *concentration*. Such a *velocity* will create a *dipole distortion* of the CBR. Setting this *velocity* equal to the observed 600 km/sec *dipole* in the CBR yields D = 80 Mpc, so our *galaxy* would be in the central 12% by distance of the *concentration*. Of course, if our *galaxy* has an additional *velocity* component relative to the *expansion* of the *concentration*, this *distance* could be greater or less than that calculated.

...

Another contribution to anisotropy is from *inhomogenities* in the *local density* of *magnetic filaments*, which is assumed to be proportional to the *local magnetic energy density*. An accurate model of this would require a detailed knowledge of magnetic field strength throughout the local universe, which is not available. However, we can make good estimates of the maximum anisotropies to be expected from this source.

Magnetic field enhancements are to be expected along the roughly cylindrical *filaments* that contain large *concentrations* of *galaxies*.

...

A final source of anisotropy is variations in the *intrinsic temperature* of the various parts of the large-scale *concentration* within cutoff distance C (around 700 Mpc). Such *temperature variations* would be expected only over large physical distances, since equilibriation would be quite efficient within the *concentration* during the time since the generation of the *cosmic background radiation* (CBR) *energy*. Indeed, even for the largest *subconcentrations* with *radius* 60 Mpc, *temperature deviations* would be of the order of e^{-10} or 5 x 10^{-5}. However, the effect of *absorption* is such that even much larger *temperature variations* will not create large *anisotropies*. For example, for a *concentration* with *radius* 60 Mpc at a *distance* of 600 Mpc, a 1% *temperature* difference, or 4% *radiation energy* difference, will only produce an observe ΔT of 7.7 x 10^{-6}. Thus, *the absorptive model does not produce anisotropies that contradict the COBE or other CBR observations.*

6 Conclusions

A model that hypothesizes the isotropization of the cosmic background radiation (CBR) *by magnetized plasma filaments in the intergalactic medium* (IGM) *thus provides a better fit to the spectrum of the CBR than Big Bang models* and fits as well the small *anisotropy* observed in the CBR. It does not require a *total absorption* of *radio radiation* from high z objects that is beyond that allowed by observation. Indeed, observations of *radio to optical flux ratios* among *quasars* (Kellerman, et al., 1989)[10] indicate a drop of *relative radio luminosity* by more than an order of magnitude between a z of 0.25 and 1.25, *completely compatible with model predictions.*

[10] Kellerman, K. I., et al. (October, 1989). VLA observations of objects in the Palomar Bright Quasar Survey. *Astrophys. J.*, 98, 1195-207; https://doi.org/10.1086/11520.

The model does require a *very large scale inhomgenity* about 700 Mpc in *radius* and with a *mass* of about 8 x 10^{18} solar masses. This is an *extremely large concentration*. It would have a gravitational orbital *velocity* at its outer surface of 7,200 km/sec, *considerably larger than that of any observed mass concentration in the universe*, which tend to have characteristic *gravitational velocities* of at most 2,000 km/sec.

Could such a large structure form? We would expect that *for a body of plasma to contract gravitationally, it must be collisional.* As shown in Lerner (1986)[11] *a body that is collisional with ion velocities equal to the gravationally orbital velocities, it must be less than a certain mass.*

[11] Lerner, E. J. (December, 1986). Magnetic Vortex Filaments, Universal Invariants and the Fundamental Constants. *IEEE Trans. Plasma Sci.*, Special Issue on Cosmic Plasma, PS-14, 6, 690-702; https://doi.org/ 10.1109/TPS.1986.4316620.

If we define *collisional* as having a mean free path of 27 R, where R is the *radius* of a spherical body, the *maximum mass* turns out to be about 1.5×10^{18} solar masses, 5.6 times smaller than the postulated *large-scale concentration* of our model.

However, this may not be as much of a problem as it appears, since if the *effective thermal velocities* of *ions* are only 2/3 the *gravitational orbital velocity*, or *thermal energy* is about one-half *gravitational binding energy*, the *entire mass* would still be *collisional*. This issue can only be resolved with more detailed modeling or possibly simulation.

In any case, the calculations in this paper clearly show that *the Big Bang is not by any means the only theory compatible with current observations of the cosmic background radiation (CBR).*

Draine, B. T., & Lazarian, A. (February, 1998). Diffuse Galactic Emission from Spinning Dust Grains.

Astrophys. J., 494, L19–L22; https://iopscience.iop.org/article/10.1086/311167/pdf.

B.T.D. acknowledges the support of NSF grants AST-9319283 and AST-9619429, and A.L. acknowledges the support of NASA grant NAG5-2858.

Princeton University Observatory, Peyton Hall, Princeton, NJ.

Received October 14, 1997.
Accepted December 9, 1997.

Abstract

Spinning interstellar dust grains produce detectable rotational emission in the 10–100 GHz frequency range. *We calculate the emission spectrum and show that this emission can account for the "anomalous" galactic background component*, which correlates with *100 μm thermal emission from dust*. *Free-free emission* cannot account for the anomalous component. Implications for *cosmic background* studies are discussed.

---

1. INTRODUCTION

Experiments to map the *cosmic background radiation* (CBR) have stimulated renewed interest in *diffuse Galactic emission*. Sensitive observations of variations in the *microwave sky brightness* have revealed a component of the 14–90 GHz *microwave* continuum which is correlated with *100 μm thermal emission from interstellar dust* (Kogut, *et al.* 1996; de Oliveira Costa, *et al.* 1997; Leitch, *et al.* 1997).

[1] Kogut, A., Banday, A. J., Bennett, C. L., Gorski, K. M., Hinshaw, G., & Reach, W. T. (1996). *Astrophys. J.*, 460, 1.

150

[2] de Oliveira-Costa, A., Kogut, A., Devlin, M. J., Netterfield, C. B., Page, L. A., & Wollack, E. J. (1997). Galactic Microwave Emission at Degree Angular Scales. *Astrophys. J.*, 482, L17-L20; https://doi.org/10.1086/310684.

[3] Leitch, E. M., Readhead, A. C. S., Pearson, T. J., & Myers, S. T. (September, 1997). An Anomalous Component of Galactic Emission. *Astrophys. J.*, 486, L23-L26; https://www.aoc.nrao.edu/~smyers/papers/1997ApJL.Leitch.486.L23-26.pdf.

The origin of this *"anomalous" emission* has been of great interest. While the observed frequency dependence appears consistent with *free-free emission* (Kogut, *et al.*,1996; Leitch, *et al.*, 1997), *it is difficult to reconcile the observed intensities with free-free emission from interstellar gas* (see § 6 below).

[*Bremsstrahlung* is *electromagnetic radiation* produced by the deceleration of a charged particle *when deflected by another charged particle, typically an electron by an atomic nucleus.* The moving particle loses kinetic energy, which is converted into radiation (i.e., photons), thus satisfying the law of conservation of energy. The term is also used to refer to the process of producing the radiation. *Bremsstrahlung* has a continuous spectrum, which becomes more intense and whose peak intensity shifts toward higher frequencies as the change of the energy of the decelerated particles increases.

Broadly speaking, *bremsstrahlung* or *braking radiation* is any radiation produced due to the acceleration (positive or negative) of a charged particle, which includes *synchrotron radiation* (i.e., *photon emission by a relativistic particle*), *cyclotron radiation* (i.e. *photon emission by a non-relativistic particle*), and the *emission of electrons and positrons during beta decay.* However, the term is frequently used in the more narrow sense of *radiation from electrons* (from whatever source) *slowing in matter.*

Bremsstrahlung emitted from plasma is sometimes referred to as *free–free radiation*. This refers to the fact that *the radiation in this case is created by electrons that are free (i.e., not in an atomic or molecular bound state) before, and remain free after, the emission of a photon.* In the same parlance, *bound–bound radiation* refers to discrete spectral lines (an electron "jumps" between two bound states), while *free–bound radiation* refers to the radiative combination process, in which a free electron recombines with an ion.]

A recent investigation of the *rotational dynamics* of very small *interstellar grains* (Draine & Lazarian, 1998, hereafter DL98)[4] leads us instead to propose that this 10–100 GHz "anomalous" component of the diffuse Galactic background is produced by *electric dipole rotational emission* from *very small dust grains* under *normal interstellar conditions*.

[4] Draine, B. T., & Lazarian, A. (1998). in preparation (DL98). [Published as Draine, B. T., & Lazarian, A. (November, 1998). Electric Dipole Radiation from Spinning Dust Grains. *Astrophys. J.*, 508, 157-79; https://doi.org/10.1086/306387. See below.]

[The *frequency* of *cosmic background radiation* (CBR) ranges from about 0.3 GHz to 630 GHz.]

Below we describe briefly our assumptions concerning the *interstellar grain* properties (§ 2), the dynamics governing the rotation of small grains (§ 3), the predicted *emission spectrum* and *intensity* (§ 4) and how it compares with *observations* (§ 5), and why the *observed emission* cannot be attributed to *free-free emission* (§ 6). In §7 we discuss implications for future *cosmic background radiation* (CBR) studies.

2. GRAIN PROPERTIES

The *observed emission* from *interstellar diffuse clouds* at 12 and 25µm (Boulanger & Perault, 1988)[5] is believed to be *thermal emission* from *grains* that are small enough that absorption of a single *photon* can raise the *grain temperature* to 150 K for the 25 µm *emission* and 300 K for the 12 µm *emission*.

[5] Boulanger, F., & Perault, M. (July, 1988). Diffuse infrared emission from the galaxy. I - Solar neighborhood. *Astrophys. J.*, 330, 964-85; https://doi.org/10.1086/166526.

Such grains contain $\sim 10^2$–10^3 *atoms* and must be sufficiently numerous to account for 20% of the *total rate of absorption of energy* from *starlight*[6].

[6] Assuming a Debye temperature Θ = 420 K, a 6 eV *photon* can heat a particle with N = 277 *atoms* to T = 200 K.

A substantial fraction of these very small *grains* may be *hydrocarbon molecules*, as indicated by *emission bands* at 6.2, 7.7, 8.6, and 11.3 µm observed from diffuse clouds by IRTS (Onaka, *et al.*, 1996)[7] and ISO (Mattila, *et al.*, 1996)[8].

[7] Onaka, T., Yamamura, I., Tanabe, T., Roellig, T., & Yuen, L. (1996). Detection of the Mid-Infrared Unidentified Bands in the Diffuse Galactic Emission by IRTS. *PASJ*, 48, L59-63.
[8] Mattila, K., Lemke, D., Haikala, L. K., Laureijs, R. J., Leger, A., Lehtinen, K., Leinert, C., & Mezger, P. G. (November, 1996). *Astronomy and Astrophysics*, 315, L353-6.

If the *grains* are primarily *carbonaceous*, they must contain 3%–10% of the *carbon* in the *interstellar medium* (in the model of Desert, Boulanger, & Puget, 1990[9], *polycyclic aromatic hydrocarbon* molecules contain 9% of the *carbon*).

[9] Desert, F. -X., Boulanger, F., & Puget, J. L. (1990), *Astronomy and Astrophysics*, 237, 215.

The size distribution is uncertain. As in DL98, for our standard model "A" we assume a *grain size distribution* consisting of a power law $dn/da \propto a^{-3.5}$ (Mathis, Rumpl, & Nordsieck, 1977[10]; Draine & Lee, 1984[11]) plus a lognormal distribution containing 5% of the *carbon*; 50% of the *mass* in the lognormal distribution (i.e., 2.5% of the C *atoms*) is in *grains* with $N < 160$ *atoms*.

[10] Mathis, J. S., Rumpl, W., & Nordsieck, K. H. (October, 1977). The size distribution of interstellar grains. *Astrophys. J.*, 217, 425-33.
[11] Draine, B. T., & Lee, H. M. (October, 1984). Optical Properties of Interstellar Graphite and Silicate Grains. *Astrophys. J.*, 285, 89.

The *grains* are assumed to be disk-shaped for $N < 120$ and spherical for $N > 120$.

The *microwave emission* from a *spinning grain* depends on the component of the *electric dipole moment* perpendicular to the *angular velocity*. Following DL98, we assume that a *neutral grain* has a *dipole moment* $\mu = N^{1/2}\beta$, where N is the *number of atoms* in the *grain*. Recognizing that there will be a range of *dipole moments* for *grains* of a given N, we assume that 25% of the *grains* have $\beta = 0.5\ \beta_0$, 50% have $\beta = \beta_0$, and 25% have $\beta = 25\ \beta_0$. For our standard models we will take $\beta_0 = 0.4$ debye, but since b is uncertain, we will examine the effects of varying β_0.

If the *grain* has a *charge Ze*, then it acquires an additional *electric dipole component* arising from the fact that the centroid of the *grain charge* will in general be displaced from the center of *mass* (Purcell, 1979)[12]; we follow DL98 in assuming a characteristic displacement of $0.1a_x$, where a_x is the rms distance of the *grain* surface from the center of *mass*, and in computing the *grain charge distribution* f(Z), including collisional *charging* and *photoelectric emission*.

[12] Purcell, E. M. (July, 1979). Suprathermal rotation of interstellar grains. *Astrophys. J.*, 231, 404-6.

Finally, we assume that the *dipole moment* is, on average, uncorrelated with the *spin axis*, so that the mean-square *dipole moment* transverse to the *spin axis* is

$$\mu^2_\perp = (2/3) \, (\beta^2 N + 0.01 \, Z^2 e^2 a_x{}^2) \qquad (1)$$

We will show results for six different *grain* models. Models B and C differ from our preferred model A by having β_0 decreased or increased by a factor of 2. Model D differs by having twice as many small *grains* in the lognormal component. In model E even the smallest *grains* are spherical, whereas in model F the smallest *grains* are assumed to be linear.

3. ROTATION OF INTERSTELLAR GRAINS

Rotation of *small interstellar grains* has been discussed previously, including the fact that the *rate of rotation* of *very small grains* would be limited by *electric dipole emission* (Draine & Salpeter, 1979)[13]. Ferrara & Dettmar (1994)[14] recently estimated the rotational emission from *very small grains*, noting that it could be observable up to 100 GHz, but they did not self consistently evaluate the *emission spectrum* including the effects of *radiative damping*.

[13] Draine, B. T., & Salpeter, E. E. (July, 1979). On the physics of dust grains in hot gas. *Astrophys. J.*, 231, 77-94.

[14] Ferrara, A., & Dettmar, R. -J. (May, 1994). Radio-emitting dust in the free electron layer of spiral galaxies: Testing the disk/halo interface. *Astrophys. J.*, 427, 155-9; https://doi.org/10.1086/174128. See above.

A number of distinct physical processes act to excite and damp rotation of *interstellar grains*, including *collisions with neutral atoms*, *collisions with ions* and *long-range interactions with passing ions* (Anderson &

Watson, 1993)[15], *rotational emission of electric dipole radiation, emission of infrared radiation, formation of H_2 molecules on the grain surface,* and *photoelectric emission.*

[15] Anderson, N., & Watson, W. D. (1993). *Astronomy and Astrophysics*, 270, 477.

We use the rates for these processes summarized in DL98; the results reported here assume that H_2 does *not* form on *small grains*, so that there is no contribution by H_2 formation to the *rotation*. As discussed in DL98, observed rates of H_2 formation on *grains* limit the H_2 formation torques so that they can make at most a minor contribution to the rotation of the *very small grains* of interest here.

Emission *in the10–100 GHz region* is dominated by *grains containing* $N \leq 10^3$ *atoms*. For these *very small grains* under "*cold neutral medium*" (CNM) conditions (see Table 2 below), *rotational excitation* is dominated by *direct collisions with ions* and "*plasma drag*". The *very smallest grains* ($N \leq 150$) have their *rotation* damped primarily by *electric dipole emission*; for $150 \leq N \leq 10^3$ *plasma drag* dominates.

4. PREDICTED EMISSION

Following DL98, we solve for the *rms rotation rate* $<v^2>^{1/2}$, which depends on the local environmental conditions (*density* n_H, *gas temperature* T, *fractional ionization* n_e/n_H, and the *intensity of the starlight background*), the *grain size*, and the assumed value of β in equation (1). Then, assuming a Maxwellian distribution of *angular velocities* for each *grain size* and β, we integrate over the *size distribution* to obtain the *emission spectrum* for the grain population.

Interstellar gas is found in a number of characteristic physical states (see, e.g., McKee 1990)[16].

[16] McKee, C. F. (1990). In ASP Conf. Ser. 12, *The Evolution of the Interstellar Medium*, ed. L. Blitz, ASP, San Francisco, CA, 3.

By *mass*, most of the *gas* and *dust* is found in the *"warm ionized medium"* (WIM), *"warm neutral medium"* (WNM), *"cold neutral medium"* (CNM), or in *molecular clouds* (MC). In the diffuse regions (WIM, WNM, CNM) *the bulk of the dust is heated by the general starlight background to temperatures* ~ 18K, where it radiates strongly at ~ 100 mm—thus the 100 mm emission traces the *mass* of the WIM, WNM, and CNM material. Relatively little *molecular material* is present at | b | < 30 °, where sensitive *cosmic background radiation* (CBR) observations are directed. Hence, we consider only the *"warm ionized medium"* (WIM), *"warm neutral medium"* (WNM), *"cold neutral medium"* (CNM) phases in the present discussion.

In Table 1 we list the assumed properties for each of the *interstellar medium components* that we consider here. For b > 30 °, 21 cm observations (Heiles, 1976[17]) indicate N_H (WNM) + N_H (CNM) ≈ 3.4 x 10^{20} (csc b − 0.2) cm^{-2}, divided approximately equally between WNM and CNM phases (Dickey, Salpeter, & Terzian, 1978[18]).

[17] Heiles, C. (March, 1976). An almost Complete Survey of 21 Centimeter Line Radiation for |b| Greater than or Equal to 10 Degrees. III. the Interdependence of H i, Galaxy Counts, Reddening, and Galactic Latitude. *Astrophys. J.*, 204, 379-402.
[18] Dickey, J. M., Salpeter, E. E., & Terzian, Y. (January, 1978). Galactic neutral hydrogen emission-absorption observations from Arecibo. *Astrophys. J., Suppl.*, 36, 77-114.

Dispersion measures toward *pulsars* in four globular clusters (Reynolds, 1991[19]) indicate an ionized component N_H (WIM) ≈ 5 x 10^{19} csc b cm^{-2}, after attributing ~ 20% of the *electrons* to hot *ionized gas* (McKee, 1990[16]).

[19] Reynolds, R. J. (May, 1991). Line integrals of n_e and (n_e)-squared at high Galactic latitude. *Astrophys. J.*, 372, L17-20.

In Figure 1 we show the predicted *rotational emissivity* per H *nucleon*, where we have taken the weighted average of emission from CNM,

WNM, and WIM, according to the *mass* fractions in Table 2. The thick curve is for our preferred *grain model* A; the results shown for *grain models* B–F serve to illustrate the uncertainty in the predicted *emission*. All models have the *grain emissivity* peaking at about ~ 30 GHz; when the *smallest grains* are taken to be spheres (model E), the *emission* peak shifts to ~ 35 GHz because of the reduced *moment of inertia* of these *smallest grains*.

In addition to the *rotational emission* from the *very small grains*, we expect *continuum emission* from the *vibrational modes* of the *larger dust grains*, which are thermally excited according to the *grain vibrational temperature* T_d. The *vibrational emission* per H *atom* is assumed to vary as $j_v/n_H = Av^\alpha B_v(T_d)$. We show the *emissivity* for α = 1.5, 1.7, and 2; in each case, we adjust A and T_d so that vj_v peaks at λ = 140 μm, with a peak *emissivity* $4\pi vj_v/n_H = 3 \times 10^{-24}$ ergs s^{-1} H^{-1} (Wright *et al.*, 1991[20]; cf. Fig. 5 of Draine, 1995[21]).

[20] Wright, E. L., *et al.* (1991). *Astrophys. J.*, 381, 20.
[21] Draine, B. T. (1995), in ASP Conf. Ser. 80, *Physics of the Interstellar Medium and Intergalactic Medium*, ed. A. Ferrara, C. F. McKee, C. Heiles, & P. R. Shapiro, San Francisco ASP, 133.

The resulting values of α and T_d are within the range found by Reach, *et al.* (1995)[22].

[22] Reach, W. T., *et al.* (September, 1995). Far-Infrared Spectral Observations of the Galaxy by COBE. *Astrophys. J.*, 451, 188-99.

We will use α = 1.7 to estimate the *thermal emissivity*; by comparison with the estimates for α = 1.5 and α = 2, we see that the *thermal emission* at ~ 100 GHz is uncertain by at least a factor of ~ 2. We expect the *thermal emission* to dominate for $v \geq 70$ GHz, with the *rotational emission* dominant at *lower frequencies*.

5. COMPARISON WITH OBSERVATIONS

Figure 2 shows the emission per H *nucleon* for our preferred parameters (model A, $\alpha = 1.7$), for each phase as well as the weighted average over the three phases. Also shown are the *observational* results of Kogut, *et al.* (1996) (based on cross correlation of COBE DMR 31.5, 53, and 90 GHz maps with COBE DIRBE 100 μm maps); de Oliveira-Costa, *et al.* (1997) (based on the cross-correlation of Saskatoon 30 and 40 GHz maps with COBE DIRBE 100 μm maps); and Leitch, *et al.* (1997) (based on cross-correlation of OVRO mapping at 14.5 and 32 GHz with IRAS 100 μm maps). *All three papers reported strong positive correlations of microwave emission with 100 mm thermal emission from dust.* The observed correlation of 100 μm *emission* with 21 cm *emission*, $I_\nu\,(100\ \mu m) \approx 0.85$ MJy sr^{-1} ($N_H/10^{20}$ cm^{-2}) (Boulanger & Perault, 1988), allows us to infer the *excess microwave emission* per H *atom*, as shown in Figure 2.

While Kogut, *et al.* attribute only 50% of the 90 GHz signal to *thermal emission, we estimate that the 90 GHz signal is predominantly vibrational emission from dust.* The *rotational emission* that we predict appears to be in good agreement with the 30–50 GHz measurements by Kogut, *et al.*, de Oliveira Costa, *et al.*, and Leitch, *et al.* *We conclude that rotational emission from small dust grains accounts for a substantial fraction of the "anomalous" Galactic emission at 30–50 GHz.*

Leitch, *et al.* also report *excess emission* at 14.5 GHz, which is correlated with 100 μm *emission from dust.* For our model A, *rotational emission from dust grains accounts for only 30% of the reported excess emission at 14.5 GHz.* Additional *emission* at 14.5 GHz could be obtained by changes in our adopted parameters: model B, in which *dipole moments* are increased by a factor of 2, would be in good agreement with the 14.5 GHz result of Leitch, *et al.* However, *the assumed dipole moment is larger than we consider likely, so we do not favor this model.* We could, of course, also improve agreement by increasing the *number of small grains* in the size range (N ≈ 100) primarily *radiating near 14.5 GHz.*

We note that there may be systematic variations in the *small grain population* from one region to another. The signal reported by Leitch, *et al.* at 30 GHz is about a factor of 3 larger than the results of Kogut, *et al.* and de Oliveira-Costa, *et al.*, both of which average over much larger areas than the observations of Leitch, *et al.* Additional observations to measure more precisely the *Galactic emission* will be of great value. We also note that although Leitch, *et al.* found no correlation of 14.5 GHz excess with synchrotron maps at 408 MHz, *synchrotron emission and rotational emission from dust are expected to contribute approximately equally to the diffuse background near 14 GHz* (see Fig. 3). We conjecture that some of the 100 μm–correlated 14.5 GHz radiation observed by Leitch, *et al. may be synchrotron emission that is enhanced by increased magnetic field strengths near concentrations of gas and dust.*

6. WHY NOT FREE-FREE EMISSION?

Based on the *observed frequency dependence* of the *excess radiation,* Kogut, *et al. proposed that it was a combination of thermal emission from dust plus free-free emission from ionized hydrogen.* Leitch, *et al.* reached the same conclusion based on their 14.5 and 32 GHz measurements.

Leitch, *et al.* noted that if the proposed spatially varying *free-free emission* originated in gas with T ≤ 10^4 K, it would be accompanied by Hα *emission* at least 60 times stronger than the *observed* variations in the Hα *sky brightness* (Gaustad, McCullough, & Van Buren, 1996[23]) on the same *angular* scales.

[23] Gaustad, J. E., McCullough, P. R., & Van Buren, D. (April, 1996). An Upper Limit on the Contribution of Galactic Free-Free Emission to the Cosmic Microwave Background near the North Celestial Pole. *PASP*, 108, 351.

Noting that the ratio of Hα to *free-free* radio continuum drops as the gas temperature is increased above ~ 3 x 10^4 K, Leitch, *et al. proposed that the "anomalous" emission originated in gas at T ≥ 10^6 K.*

However, this proposal appears to be untenable. According to Leitch, *et al.*, the *observed emission excess* would require an *emission* measure with EM ≈ 130 $T_6^{0.4}$ cm^{-6} pc with $T_6 \equiv$ (T/10^6 K) ≥ 1 near the North Celestial Pole (NCP; l ≈ 123 °; b ≈ 27.4 °). Kogut, *et al.* and de Oliveira-Costa, *et al.* reported by a weaker signal at 30 GHz, corresponding to an *emission* measure ~ 40 $T_6^{0.4}$ cm^{-6} pc toward the NCP. The DMR observations (Kogut, *et al.*, 1996) cover a substantial fraction of the sky and would imply an *emission* measure *normal to the Galactic disk* ~ 2 sin 27 ° x 40 $T_6^{0.4}$ cm^{-6} pc. For T ≥ 10^6 K, the radiative cooling rate can be approximated by ~ 1 x 10^{-22} (T_6^{-1} + 0.3 $T_6^{1/2}$) $n_H\,n_e$ ergs cm^3 s^{-1} (cf. Bohringer & Hensler, 1989[24]) so the *power* radiated per disk area would be ~ 30($T_6^{-0.6}$ + 0.3 $T_6^{0.9}$) L$_\odot$ pc^{-2}, far in excess of the 0.3 L$_\odot$ pc^{-2} *energy input* for one 10^{51} ergs *supernova* per 100 yr per 10 kpc *radius disk.*

[24] Bohringer, H., & Hensler, G. (May, 1989). Metallicity-dependence of radiative cooling in optically thin, hot plasmas. *Astronomy and Astrophysics*, 215, 1, 147-9.

Evidently, *the proposed attribution of the "anomalous emission" to bremsstrahlung from hot gas can be rejected on energetic grounds.*

7. DISCUSSION

Very small dust grains have been invoked previously to explain the observed 3–60 μm *emission* from *interstellar dust* (Leger & Puget, 1984[25]; Draine & Anderson, 1985[26]; Desert, Boulanger, & Puget, 1990[9]).

[25] Leger, A., & Puget, J. L. (August, 1984). Identification of the 'unidentified' IR emission features of interstellar dust? *Astronomy and Astrophysics*, 137, 1, L5-8.

[26] Draine, B. T., & Anderson, N. (May, 1985). Temperature fluctuations and infrared emission from interstellar grains. *Astrophys. J.*, 292, 494-9.

We have calculated the spectrum of rotational emission expected from such dust grains and shown that it should be detectable in the 10–100 GHz region. In fact, we argue that this emission has already been detected by Kogut, *et al.* (1996), de Oliveira-Costa, *et al.* (1997), and Leitch, *et al.* (1997).

Rotational emission from very small grains and "thermal" (i.e., vibrational) emission by larger dust grains is an important foreground that will have to be subtracted in future sensitive experiments to measure angular structure in the cosmic background radiation (CBR). To illustrate the relative importance of the emission from dust, in Figure 3 we show the estimated rms variations in *Galactic emission* near the NCP ($l \approx 123°$; $b \approx 27.4°$), representative of intermediate Galactic latitudes. The H I column density is taken to be $N(H^0) = 6.05 \times 10^{20}$ cm^{-2} (Hartmann & Burton, 1997[27]), plus an additional *column density* $N(H^+)$ $= \sim 1.3 \times 10^{20}$ cm^{-2} of the WIM. We take 20% as an estimate of the rms variations in *column density* on *angular* scales of a few degrees.

[27] Hartmann, D., & Burton, W. B. (1997). *Atlas of Galactic Neutral Hydrogen*. Cambridge Univ. Press, Cambridge.

We also show the *synchrotron background* near the North Celestial Pole (NCP), based on a total antenna temperature of 3.55 K at 1.42 GHz (Reich, 1982[28]) and $Iv \propto v^{-1}$ at higher *frequencies*. The *synchrotron background* is smoother than the H I; we take 5% as an estimate of the rms *synchrotron* variations on *angular* scales of a few degrees.

[28] Reich, W. (May, 1982). A radio continuum survey of the northern sky at 1420 MHz. I. *Astronomy and Astrophysics*, Suppl., 48, 219-97.

From the standpoint of minimizing confusion with non-CBR backgrounds, *70–100 GHz appears to be the optimal frequency window.*

The "thermal" emission originates in larger grains, which are known to be partially aligned with their long axes perpendicular to the local magnetic field; above 100 GHz, most of the emission is thermal, and we estimate that this will be ~ 5% linearly polarized perpendicular to the *magnetic field* (Dotson, 1996[29] has observed up to 7% polarization at 100 μm toward M17).

[29] Dotson, J. L. (1996). Polarization of the Far-Infrared Emission from M17. *Astrophys. J.*, 470, 566.

Below ~ 50 GHz, most of the emission is rotational. At this time, it is not clear whether the *angular momentum* vectors of *very small dust grains* will be aligned with the *Galactic magnetic field*. Ultraviolet observations of *interstellar polarization* place limits on such alignment (Kim & Martin, 1995[30]) but do not require it to be zero.

[30] Kim, S. H., & Martin, P. G. (1995). *Astrophys. J.*, 444, 293.

If the *grain angular momenta* tend to be aligned with the *Galactic magnetic field*, then the *rotational emission* will tend to be polarized perpendicular to the *magnetic field*. Physical processes that could produce alignment of these *small grains* will be discussed in future work (Draine & Lazarian, 1998[31]).

[31] Draine, B. T., & Lazarian, A. (November, 1998). Electric Dipole Radiation from Spinning Dust Grains. *Astrophys. J.*, 508, 157-79. See below.

Even modest (~ 1%) polarization of the dust emission could interfere with efforts to measure the small polarization of the *cosmic background radiation* (CBR) introduced by cosmological *density* fluctuations.

According to our model, the ~ 30 GHz emission and diffuse 12 μm emission should be correlated, since both originate in grains containing ~ 10^2 atoms; future satellite observations of the diffuse background can test this. Future observations by both ground-based experiments and satellites such as MAP and PLANCK will be able to characterize more precisely the *intensity* and *spectrum* of *emission* from *interstellar dust grains* in the 10–200 GHz region. Measurements of the 10–100 GHz emission will constrain the abundance of *very small grains*, including spatial variations, as well as their possible alignment.

Draine, B. T., & Lazarian, A. (November, 1998). Electric Dipole Radiation from Spinning Dust Grains.

Astrophys. J., 508, 157-79; https://doi.org/10.1086/306387. Also at https://iopscience.iop.org/article/10.1086/306387/pdf.

Princeton University Observatory, Peyton Hall, Princeton, NJ 08544.

Received February 18, 1998.
Accepted June 25, 1998.

Abstract

We discuss the rotational excitation of small interstellar grains and the resulting electric dipole radiation from spinning dust. Attention is given to *excitation* and *damping* of *grain rotation* by *collisions with neutrals*, *collisions with ions*, *"plasma drag"*, *emission of infrared radiation*, *emission of electric dipole radiation*, *photoelectric emission*, and *formation of H_2 on the grain surface*. Electrostatic "focusing" can substantially enhance the rate of rotational excitation of *grains* colliding with *ions*. Under some conditions, *"plasma drag"* - due to interaction of the *electric dipole moment of the grain with the electric field produced by passing ions – dominates both rotational damping and rotational excitation. Emissivities* are estimated for *dust* in different phases of the *interstellar medium*, including *diffuse H_I clouds, warm H_I, low-density photo-ionized gas*, and *cold molecular gas. Spinning dust grains could explain much, and perhaps all, of the 14-50 GHz background component recently observed by Kogut, et al., de Oliveira-Costa, et al., and Leitch, et al.* Future sensitive measurements of *angular* structure in the *microwave sky brightness* from the ground and from space should detect this *emission* from *high latitude H_I clouds*. It should be possible to detect rotational emission from *small grains* by ground based pointed observations of molecular clouds, unless these *grains* are less abundant there than is currently believed.

165

1. Introduction

Experiments to map the *cosmic microwave background* radiation have stimulated renewed interest in *diffuse Galactic emission*. Most recently, Kogut, *et al.* (1996)[1], de Oliveira-Costa, *et al.* (1997)[2] and Leitch, *et al.* (1997)[3] have reported a new component of *galactic microwave emission* which is correlated with *100μm thermal emission from interstellar dust*.

[1] Kogut, A., Banday, A. J., Bennett, C. L., Gorski, K. M., Hinshaw, G., & Reach, W. T. (1996). *Astrophys. J.*, 460, 1.

[2] de Oliveira-Costa, A., Kogut, A., Devlin, M. J., Netterfield, C. B., Page, L. A., & Wollack, E. J. (1997). Galactic Microwave Emission at Degree Angular Scales. *Astrophys. J.*, 482, L17-L20; https://doi.org/10.1086/310684.

[3] Leitch, E. M., Readhead, A. C. S., Pearson, T. J., & Myers, S. T. (September, 1997). An Anomalous Component of Galactic Emission. *Astrophys. J.*, 486, L23-L26; https://www.aoc.nrao.edu/~smyers/papers/1997ApJL.Leitch.486.L23-26.pdf.

Kogut, *et al.*[1] found the emission excess to have I_v(31.5 GHz) ≈ I_v(53 GHz), and Leitch, *et al.*[3] found I_v(14.5 GHz) ≈ I_v(32 GHz), consistent with the spectrum of *free-free emission*, but non-detection of Hα emission in these directions is inconsistent with *free-free emission* accounting for the microwave excess unless the *plasma* temperature T > ~ 106 K. Leitch, *et al.*[3] therefore proposed that the *observed emission* was *free-free emission* from shock-heated gas in a *supernova* remnant.

[*Bremsstrahlung emitted from plasma* is sometimes referred to as *free–free radiation*. This refers to the fact that *the radiation in this case is created by electrons that are free (i.e., not in an atomic or molecular bound state) before, and remain free after, the emission of a photon*. In the same parlance, *bound–bound radiation* refers to discrete spectral lines (an *electron* "jumps" between two bound states), while *free–bound radiation* refers to the radiative combination process, in which a *free electron* recombines with an *ion*.

Bremsstrahlung is *electromagnetic radiation* produced by the deceleration of a charged particle *when deflected by another charged particle, typically an electron by an atomic nucleus*. The moving particle loses kinetic energy, which is converted into radiation (i.e., photons), thus satisfying the law of conservation of energy. The term is also used to refer to the process of producing the radiation. *Bremsstrahlung* has a continuous spectrum, which becomes more intense and whose peak intensity shifts toward higher frequencies as the change of the energy of the decelerated particles increases.]

Draine & Lazarian (1998, hereafter DL98[4]) showed, however, that the *observed microwave excess* could not be due to *free-free emission* from *hot gas*, as this would require an energy injection rate at least 2 orders of magnitude greater than that the energy input due to *supernovae*. DL98 showed that the *microwave excess* could in fact be *electric dipole emission* from *rapidly-rotating dust grains*.

[4] Draine, B. T., & Lazarian, A. (February, 1998). Diffuse Galactic Emission from Spinning Dust Grains. *Astrophys. J.*, 494, L19–L22; https://iopscience.iop.org/article/10.1086/ 311167/pdf. See above.

To predict the *intensity* of *dipole emission* one needs (1) the *numbers of small grains*, (2) their *dipole moments*, and (3) their *rotational velocities*. The observed *intensity* of 12 and 25µm *emission* from *interstellar clouds* allows us to estimate the numbers of *very small grains* (Leger & Puget, 1984[5]; Draine & Anderson, 1985[6]; Desert, Boulanger, & Puget, 1990[7]).

[5] Leger, A., & Puget, J. L. (August, 1984). Identification of the 'unidentified' IR emission features of interstellar dust? *Astronomy and Astrophysics*, 137, 1, L5-8.

[6] Draine, B. T., & Anderson, N. (May, 1985). Temperature fluctuations and infrared emission from interstellar grains. *Astrophys. J.*, 292, 494.

[7] Desert, F. -X., Boulanger, F., & Puget, J. L. (1990), *Astronomy and Astrophysics*, 237, 215.

To estimate the *dipole moments*, we consider the likely displacements between the *charge* and *mass* centroids for *grains*, plus the *intrinsic dipole moment* arising from *polarized chemical bonds* in the *grain material*. However, the main thrust of the present paper is a comprehensive study of the *rotational dynamics of small grains*, in order to estimate their rotation rates.

Ferrara & Dettmar (1994)[8] showed that *small grains* rotating with *thermal rotation rates* could produce detectable *radio emission*, but they did not address the details of *rotational excitation* and *damping*.

[8] Ferrara, A., & Dettmar, R. -J. (May, 1994). Radio-emitting dust in the free electron layer of spiral galaxies: Testing the disk/halo interface. *Astrophys. J.*, 427, 155-9; https://doi.org/10.1086/174128. See above.

Rotational excitation of *small grains* (= *large molecules*) has been discussed previously by Rouan, *et al.*, (1992)[9], who described the effects of *collisions with gas atoms* and *absorption and emission of radiation*.

[9] Rouan, *et al.* (January, 1992). Physics of the rotation of a PAH molecule in interstellar environments. *Astronomy and Astrophysics*, 253, 498-514.

In the present work we reexamine this problem, and include the important effects of collisions with ions, which were neglected in the study by Rouan, *et al.*[9] We also include the effects of "*plasma* drag" due to interaction of the grain with passing *ions*, first considered by Anderson & Watson, (1993)[10].

[10] Anderson, N., & Watson, W. D. (1993). *Astronomy and Astrophysics*, 270, 477.

We derive rates for *rotational damping* and *excitation* due to *plasma* drag which are somewhat larger than estimated by Anderson & Watson, (1993)[10].

We discuss rotational excitation as a function of both grain size and environmental conditions. Electrostatic focussing of *ions* makes them very effective at delivering *angular momentum* to the *grains*; *ion collisions* can dominate the *rotational excitation* of *very small grains* even in predominantly neutral regions. In addition, coupling of the *plasma* to the *grain dipole moment* can contribute significantly to *rotational damping* and *excitation*. Within regions of high *photon* density, *emission* of *infrared photons* can be extremely important.

For our adopted population of *very small grains*, we predict the *microwave emissivity* of various phases of the *interstellar medium*, ranging from *diffuse gas* to *molecular clouds*. We expect detectable levels of *emission* from *spinning dust grains* in all of these phases.

The paper is organized as follows: After describing the environments which we consider (§2), we discuss the *electric dipole moments* expected for *small dust grains* (§3). The various *rotational damping* processes are reviewed in §4, and the *rotational excitation mechanisms* in §5. Using these rates for *rotational excitation* and *damping*, we calculate the resulting *rate of grain rotation* in §6. The importance of the impulsive nature of the *rotational excitation* is discussed in §7, and the effects of *centrifugal stresses* in §8.

The reader interested primarily in the *predicted emission* may wish to skip §§2–8 and proceed directly to §9, where we describe our assumptions concerning the *grain size distribution* in various environments. We present the *microwave emission spectra* expected for *grains* in both *diffuse regions* and *molecular clouds* in §10. The detection of this *microwave emission* from pointed observations of dense clouds is discussed in §11. The principal uncertainties in our estimates are discussed in §12, and our results are summarized in §13.

2. Environments

We will consider *grains* in five different idealized "phases" of the *interstellar medium*: "*Cold Neutral Medium*" (CNM), "*Warm Neutral Medium*" (WNM), "*Warm Ionized Medium*" (WIM), "*Molecular Cloud*" (MC), and "*Dark Cloud*" (DC). In Table 1 we give the adopted values of nH (H *nucleon density*), T (*gas kinetic temperature*), χ (*starlight intensity* relative to the average starlight background), and other properties of these phases.

...

3. Grain Properties

...

13. Discussion

We have presented above a detailed study of rotational excitation and damping for grains with sizes less than $\sim 10^{-6}$ cm. In addition to processes included in earlier work by Rouan, *et al.* (1992)[9], we have included both direct collisions with ions and "*plasma* drag".

Ion collisions and *plasma* drag, together with damping by infrared and microwave emission, dominate rotational excitation and damping for most interstellar environments. It is therefore important to include these processes in other studies of rotational excitation, such as those proposing to explain features of the diffuse interstellar bands (Rouan, Leger, & Coupanec, 1997[11]).

[11] Rouan, D, Léger, A., & Le Coupanec, P. (1997). Carrier of the λ 5797 DIB in the ISM and in the Red Rectangle: a suprathermally rotating PAH molecule? *Astronomy & Astrophysics,* 324, 661-73; https://cdsarc.u-strasbg.fr/ftp/vizier/aa/papers/7324002/2300661/small.htm.

For the assumed distribution of *grain sizes* and *electric dipole moments*, we predict *microwave emission* from spinning dust grains which can account for the "anomalous emission" observed recently in the 15–90

170

GHz range. The predicted *intensities* are uncertain, however, as they depend on three poorly-known factors:

1. *The abundances and size distribution of the small grains.* This uncertainty is greatest in the case of dense/dark regions, where the absence of starlight denies us the evidence (12– 25 μm emission) which requires an abundance of *ultrasmall grains* in diffuse regions.
2. *The charge distribution of the small grains*, which affects the rates of *rotational excitation* and *damping*. We have used standard estimates for photoelectric cross sections (Bakes & Tielens, 1994[12]) and *electron capture cross sections* (Draine & Sutin, 1987) for *very small grains*.

> [12] Bakes, E. L. O., & Tielens, A. G. G. M., (June, 1994). The Photoelectric Heating Mechanism for Very Small Graphitic Grains and Polycyclic Aromatic Hydrocarbons. *Astrophys. J.*, 427, 822; https://doi.org/10.1086/174188.
>
> [13] Draine, B. T., & Sutin, B. (September, 1987). Collisional Charging of Interstellar Grains. Astrophys. J., 320, 803; https://doi.org/10.1086/165596.

The uncertainties in these quantities affect both the electric dipole moment and, more importantly, the rate of angular momentum exchange with ions.
3. *The electric dipole moments of both neutral and charged small grains.*

The shape of the small grains (chainlike vs. sheetlike vs. quasi-spherical) is also uncertain, but is less critical than the above factors. A fifth factor – the efficiency of H_2 formation on *small grains* – is not critical for these estimates: in Fig. 12 only the grains in the "*Cold Neutral Medium*" (CNM) show appreciable sensitivity to whether or not H_2 formation takes place, and even for the CNM component the emissivity at $v > \sim 2$ GHz is only slightly changed (see Fig. 17).

Because of these uncertainties, definitive predictions for the *rotational emission spectrum* are not yet possible. Nevertheless, within the existing theoretical and observational uncertainties *it appears that much or all of the observed 15–90 GHz "anomalous" emission is due to spinning dust grains.* The largest discrepancy between observation and theory is at 14.5 GHz, where Leitch, *et al.*[3] report *emission* about 3.5 times stronger than the *emission* predicted in Fig. 15. Additional measurements at $v < \sim 20$ GHz will be of great value to clarify whether this *emission* has another origin, or whether some of our assumptions concerning the *dust* must be modified.

Emission from rotating grains must be allowed for in studies of the cosmic microwave background radiation ... . It appears that the *small* rotating grains may be partially aligned with the *local magnetic field* (Draine & Lazarian, 1998[4]), so that the *electric dipole radiation* will be *linearly polarized*; this may present a problem for interpretation of the *cosmic microwave background* (CMB) polarization measurements by the MAP mission.

13. Summary

The principal results of this paper are as follows:

1. *Even neutral dust grains are expected to usually have electric dipole moments arising from polarized chemical bonds within the grain. Charged grains have an additional contribution to the dipole moment due to displacement of the charge centroid from the mass centroid.* Our estimate for the *dipole moment* is given by eq. (11).

2. *The excitation and damping of rotation in small grains is determined by collisions with ions and neutrals, "plasma* drag*", emission of infrared and microwave radiation, and formation of H_2 on the grain surface. Ion collisions* and *plasma* drag, omitted in previous estimates of *rotation rates*, are included in the present analysis and found to often *dominate rotational excitation and damping. Induced-dipole attraction of neutrals by charged grains, and of ions by neutral grains, can also be significant.* Because the *charge state* of the *grain*, and the fractional ionization of the gas, do not reflect thermodynamic equilibrium, *the fluctuation-dissipation theorem does not directly apply.*

172

3. *For very small grains* (a < ~ 7×10^{-8} cm), the *angular momentum of colliding ions is large compared to the r.m.s. angular momentum of the grain*, and therefore *the grain rotation history consists of "rotational spikes" separated by intervals of gradual rotational damping.*

4. *The estimated grain rotation rates are such that the small grains which have been postulated to explain the near-infrared emission feature can account for the emission observed* at 30– 50 GHz by Kogut et al., de Oliveira-Costa, *et al.*[2], and Leitch, *et al.*[3] (1997).

5. *The emission observed at 14.5 GHz by Leitch, et al.*[3] *is stronger than we estimate for spinning grains by a factor* ~ *4*. Additional determinations of emission from *dust* at frequencies < ~ 20GHz will be of great value.

6. *We predict that dark clouds should produce detectable microwave emission*, with *antenna temperatures of* > ~ 1 mK at ~ 10–30 GHz.

7. *It may be possible to detect 10-100 GHz emission from spinning grains in photon-dominated regions (PDRs), although free-free emission may dominate at these frequencies* unless the exciting *star* is quite cool.

Wright, E. L. (October, 2003.) Errors in the "The Big Bang Never Happened"
1. **Errors in Lerner's Criticism of the Big Bang**
2. **Errors in Lerner's Alternative to the Big Bang**
3. **Miscellaneous Errors.**

https://www.astro.ucla.edu/~wright/lerner_errors.html.

[The comments in this and the following article relate to the 1992 book "*The Big Bang Never Happened*", not to the online article below which was last modified in 2022.

Physical cosmologists who have commented on the book have generally dismissed it. In particular, American astrophysicist and cosmologist Edward L. Wright criticized Lerner for making errors of fact and interpretation, arguing that:
1. *Lerner's alternative model for Hubble's law is dynamically unstable*
2. *the number density of distant radio sources falsifies Lerner's explanation for the cosmic microwave background (CMB)*
3. *Lerner's explanation that the helium abundance is due to stellar nucleosynthesis fails because of the small observed abundance of heavier elements*
Lerner disputed Wright's critique. (See below.)

Wright is an American astrophysicist and cosmologist. He has worked on space missions including the Cosmic Background Explorer (COBE), Wide-field Infrared Survey Explorer (WISE), and Wilkinson Microwave Anisotropy Probe (WMAP) projects. Wright received his PhD (Astronomy in 1976) in high-altitude rocket measurement of *cosmic microwave background* (CMB) radiation from Harvard University, where he was a junior fellow. After teaching as a tenured associate professor in the MIT Physics Department for a while, Wright has been a professor at UCLA since 1981.]

Eric Lerner starts his book "*The Big Bang Never Happened*" (hereafter BBNH)[1] with the "errors" that he thinks invalidate the Big Bang.

[1] Lerner, E. J. (October, 1992). *The Big Bang Never Happened—A Reassessment of the Galactic Origin of Light Elements (GOLE) Hypothesis and its Implications.* Knopf Doubleday Publishing Group. See above.

These are
1. *The existence of superclusters of galaxies and structures like the "Great Wall" which would take too long to form from the "perfectly homogeneous" Big Bang.*
2. *The need for dark matter and observations showing no dark matter.*
3. *The Far Infrared Absolute Spectrophotometer (FIRAS) cosmic microwave background (CMB) spectrum is a "too perfect" blackbody.*

Are these criticisms correct? No, and they were known to be incorrect in 1991 when Lerner wrote his book.

[None of these comments appear in the online article *The Big Bang Never Happened* below, which was last modified in 2022.]

Let us look at the *superclusters* first.

[(1) *The existence of superclusters of galaxies and structures like the "Great Wall" which would take too long to form from the "perfectly homogeneous" Big Bang.*]

Lerner gives the example of *filaments* or *sheets* 150 million light years apart in Figure 1.1, and then asserts that material would have to travel 270 million light years to make the *structure*. Obviously 75 million light years would do the trick. With material traveling at 1000 km/sec, that would take 22.5 billion years, which is about twice as long as the probable age of the Universe. *But when the Universe was younger, everything was closer together*, so a small motion made early in the history of the Universe counts for much more than a motion made later. Thus, it was easier for the material to clump together early in the history of the Universe. ...

[*Assumes the Big Bang theory or an expanding universe.*]

Furthermore, *velocities* relative to the *Hubble flow* naturally decrease with time, so the 1000 km/sec velocity was larger in the past. ...

[*Assumes an expanding universe.*]

...

... So Lerner's "structures that take too long to grow" are just more evidence for a large amount of *dark matter*.

[*Assumes existence of dark matter.*]

...

[Lerner's response to this (below) was "There are two errors here. Even calculations by advocates of the *Big Bang* show that *the structures we observe would take about 5 times as long as the Hubble time (the hypothetical time since the Big Bang) to form, even with dark matter.* And, second, *there is no evidence that dark matter exists.*"]

[(2) The need for dark matter and observations showing no dark matter.]

So where was the "crisis"? *The "crisis" only arises if there is no dark matter.* Without *dark matter* you need 10 times larger initial perturbations and thus a 10 times larger RMS *quadrupole*, which was finally ruled out in 1991 [???] after Lerner wrote his book.

...

Is there dark matter?

There is certainly lots of evidence for *dark matter*. When one looks at *cluster of galaxies*, the *gravitational* effects of the *cluster* can be measured three ways. *One is by the orbital motions of the galaxies in the cluster*. This was first done by Zwicky in 1933[2]!

> [2] Zwicky, F. (1933). The redshift of extragalactic nebulae. *Helv. Phys. Acta.*, 6, 110-127; https://arxiv.org/abs/1612.00805; https://ned.ipac.caltech.edu/level5/March17/Zwicky/translation.pdf.

A second looks at the hot gas trapped in many big clusters of galaxies. The third way looks at the bending of light from galaxies behind the cluster by the mass in the cluster (gravitational lensing). All three methods give masses that appear to be very much larger than the mass of the stars in the galaxies in the cluster. This is usually given as the *mass-to-light ratio*, and M/L is several hundred solar units for *clusters of galaxies* and only about 3 for the *stars in the Milky Way near the Sun.* ...

The only way to satisfy these observations without a lot of dark matter is to hypothesize that the force of gravity is much stronger at large distances than Newton (or Einstein) would predict. This model is called *Modification of Newtonian Dynamics* (MOND), and it has some adherents. But *no good relativistic version of MOND exists*, and *the existence of gravitational lensing in cluster of galaxies requires a relativistic theory that makes the same change for light and for slow moving objects like galaxies.* ...

> [*Claims that only alternative to dark matter must be relativistic ???*]

...

> [**Lerner's response (below)** was "This *cold dark matter* (CDM), was hypothesized as essential for the *Big Bang* theory back in 1980--23 years ago. Since then, physicists have searched diligently with dozens of experiments for any evidence of the

177

existence of these dark matter particles here on Earth. Oddly enough *every one of the experiments has had negative results."*]

[(3) *The Far Infrared Absolute Spectrophotometer (FIRAS) cosmic microwave background (CMB) spectrum is a "too perfect" blackbody.*]

Is the cosmic microwave background (CMB) spectrum "too perfect"?

Lerner claims that *the cosmic microwave background* (CMB) *spectrum* presented by Mather in 1990 *was "too perfect"*, and that it made it impossible for large scale structure to be formed.

However, the perfect fit to the *blackbody* only ruled out *explosive structure formation* scenarios like the Ostriker and Cowie model[3].

> [3] Ostriker, J. P. & Cowie, L. L. (February, 1981). Galaxy Formation in an Intergalactic Medium Dominated by Explosions. *Astrophys. J. Lett.*, 243, L127-31.

The limits on distortion of the *cosmic microwave background* (CMB) spectrum away from a *blackbody* are now about 100 times better, and these tighter limits are easily met by models *which form large scale structure by gravitational perturbations acting on dark matter*. Models which act via *electromagnetic interactions*, like the *explosive structure formation scenario* or the *plasma Universe* have a much harder time meeting the constraints imposed by the *Far Infrared Absolute Spectrophotometer* (FIRAS) observations of the *cosmic microwave background* (CMB) spectrum. [???]

What alternative does Lerner give for the *Big Bang*? Since the *Big Bang* is based on
1. the *redshift of galaxies*
2. the *blackbody microwave background*
3. the *abundance of the light elements*.

Lerner should give alternative explanations for these three observed phenomena. What are his alternatives?

Lerner's model for the redshift

> [*A better explanation is provided by Zwicky's tired-light theory as elaborated by Compton*. See Underwood, T. G. (November, 2024). *Cosmological Redshift of Light*.]

In the BBNH, Lerner presents the Alfven-Klein model which explains the *redshift* using a portion of the Universe that starts to collapse, then the collapse is reversed. This model requires *new physics* to generate the force necessary reverse the collapse. Figure 6.2 of BBNH shows the collapse, reversal, and later expansion of a region of space.

> [There is no Figure 6.2 in the online article *The Big Bang Never Happened* below, which was last modified in 2022.]

The figure below shows space-time diagrams based on this idea. In a space-time diagram, time is plotted going upward, with the bottom being the distant past. The black lines show the paths of different clumps of *matter* (*galaxies*) as function of time. These are called "*world-lines*". The red lines show the position of light rays that reach us now at the top center of the diagrams. These are called "*light cones*". *Lerner says that only a small region of space collapsed: only a few hundred million light-years across*. This is shown on the left. *But if this were the case, then the distant galaxy at G would have a recession velocity smaller than the recession velocity of the nearby galaxy A. This is not what we observe.* Thus, a much larger region must have collapsed. This is shown on the right. Now G has a larger recession velocity than A which matches the observations.

What causes the reversal from collapse to re-expansion? Lerner claims that it is the pressure caused by the annihilation of matter and antimatter during the collapse. The green ellipse shows this high-pressure region. But only pressure differences cause forces. A pressure

179

gradient is needed to generate an acceleration. In the case of a large region of collapse, which is needed to match the observations, a larger acceleration requires a larger pressure gradient, and this gradient exists over a larger distance, leading to a greatly increased pressure.

But in relativity pressure has "weight" and causes stronger gravitational attraction. This can be seen using work W = PdV, so the *pressure* is similar to an *energy density*. Then through $E = mc^2$, this energy density is similar to a *mass density*. *If the collapsing region is big enough to match the observations, then the pressure must be so large that a black hole forms and the region does not re-expand.* Peebles discusses this problem with the *plasma* cosmology in his book "*Principles of Physical Cosmology*"[4].

[4] Peebles, P. J. E. (2020). *Principles of Physical Cosmology*. Princton University Press.

Remarkably, Lerner now disowns the *Alfven-Klein model* which plays such a big part in the BBNH, and wants me to give the proper attribution! He points out that he listed problems with the *Alfven-Klein model* in the Appendix of BBNH, but these were rather minor problems compared to the fact that it just won't work!

[There is no Appendix in the online article *The Big Bang Never Happened* below, which was last modified in 2022.]

If the *Alfven-Klein model* doesn't work, *Lerner's fall back is tired-light, which is another total failure.*

[Not true. As mentioned above, Zwicky's tired-light theory as elaborated by Compton, provides a far better explanation. See Underwood, T. G. (November, 2024). *Cosmological Redshift of Light*.]

Lerner's model for the microwave background

Lerner's model for the *cosmic microwave background* (CMB) claims that *the intergalactic medium is a strong absorber of radio waves*. His evidence for this is presented in Figure 6.19 of BBNH, which allegedly shows a decrease in the *radio to infrared luminosity ratio* as function of *distance*.

> [There is no Figure 6.19 in the online article *The Big Bang Never Happened* below, which was last modified in 2022.]

> This *absorption* is supposed to occur in *narrow filaments*, with tiny holes scattered about randomly so that distant compact radio sources like QSOs can be seen through the holes.

The best evidence against this model in also in BBNH, in Figure 6.17.

> [There is no Figure 6.17 in the online article *The Big Bang Never Happened* below, which was last modified in 2022.]

> This is a picture of *Cygnus A*, which is the *brightest extragalactic radio source*. It has a *redshift* $z = 0.056$ and is 700 million light years away, using $H_0 = 75$ as in Lerner's *Astrophys. J.* article, and looking at Figure 6.19 of BBNH, we see that it should be more than 99% absorbed. So more than 99% of the area should be blacked out by *absorbing filaments* in Figure 6.17, *but none can be seen. Cygnus A* could be plotted on Figure 6.19, but it would be off scale in the upper right corner, *completely orthogonal to Lerner's claimed trend.*

> [*Lerner's response (below)* under "Wright argues …" in section headed "A few points on Wright's misunderstanding of the *plasma* theory of the CMB" was "*Wright argues that extended radio sources contradict the absorption of radio waves by filaments in the intergalactic medium*. He points to *Cygnus A* and says that no absorbing filaments can be seen. This indicates

Wright has not read the relevant papers, which make it clear that the *absorbing filaments are quite small by astronomical standards*. Except for an initial 1987 paper, where the idea was worked out only in rough way, my elaboration of the *hypothesis of absorbing magnetic filaments* have made clear that *the filaments in general are too small to be observed directly*."]

Lerner has denied the existence of extended *high redshift radio sources*, which is pretty silly since *Cygnus A* obviously counts as one. A three times more distant extended radio source is in *Abell 2218*, with a size of 120" and a redshift of z = 0.174. Clearly this is beyond Lerner's *metagalaxy but there is no big hole in the cosmic microwave background* (CMB) *there*. The field has been studied extensively for the *Sunyaev-Zeldovich effect* and the deficit is less than a milliKelvin.

The *3CRR Atlas* has images of many distant *radio sources* with large angular size. The largest *angular size* for those sources with z > 0.4 is 3C457 which has an *angular size* of 205" and a *redshift* of z = 0.428. 7 out of the 10 sources with 0.4 < z < 0.5 in this list have sizes greater than 30". A single 30" hole in the *absorbing curtain* would have appeared as a -2 mK *anisotropy* in the Saskatoon data and *nothing like this was seen*.

Thus, *radio sources with large angular size are seen to great distances and Lerner's local absorbing curtain does not exist*.

A second objection to Lerner's local absorbing curtain is that its *density* falls inversely with *distance* from the *local density peak*, which Lerner takes to be the *Virgo supercluster*.

[***Lerner's response (below)*** in section headed "A few points on Wright's misunderstanding of the *plasma* theory of the CMB" was "*Wright's second objection, that a fractal inhomogenous collection of absorbers would lead to a non-isotropic distortion of radio sources is simply mathematically wrong*. Fractal distributions are inhomogeneous in three-space, but their projection on to 2-space, the sky, tend to be isotropic.

However, we would expect some fairly small variations in the *Cosmic Background Radiation* (CBR) because of the inhomogenous *Intergalactic Medium* (IGM) -- *where there is more density of matter, we would expect a slightly brighter CBR.* This would only be slight, because scattering and the contribution of the IGM along the same line of sight but at different distance would greatly reduce anisotropies, as described in Lerner, (May, 1995)[5].

[5] Lerner, E. J. (May, 1995). Intergalactic radio absorption and the COBE data. *Astrophys. Space Sci.*, 227, 61-81. https://doi.org/10.1007/ BF00678067.

This is what is found. There is indeed a slight correlation between *galaxy density* and *Cosmic Background Radiation* (CBR) intensity, as expected. What is particularly interesting is that *this correlation extends over all angular scales, as would be expected from the plasma viewpoint*. But in the *Black Body* hypothesis, which assumes the CBR originated BEHIND *all clusters of galaxies* and other *very dense concentrations of matter, interactions with electrons will decrease the CBR luminosity*. So, there should be an anti-correlation of galaxies and CBR on small angular scales. *Just the opposite is observed.*"]

But if the density of the *absorbers* peaks at the Virgo, *then there will be much more absorption in that direction than in the opposite direction.* This would make the distribution of radio sources on the sky very anisotropic. *But the radio sources are evenly distributed to within a few percent, so Lerner's local absorbing curtain does not exist.*

A third objection to Lerner's local absorbing curtain is that by making distant radio sources fainter, it would change the number vs flux law for radio sources in a way that is not observed.

> [**Lerner's response (below)** in section headed "A few points on Wright's misunderstanding of the *plasma* theory of the CMB" was "*Wright's third objection* illustrates the essential sloppiness of *Big Bang* thinking. He claims that statistics of flux vs number counts contradict the *absorption hypothesis. In fact, they confirm it.* Contrary to Wright's claims that $N \sim F^{-1.8}$, where N is the number of sources brighter than F, *the actual distribution is quite different.* Wright's formula is roughly true ONLY for the very brightest sources, those stronger than about 200 mJy. *For sources dimmer than that,* the relationship is very close to $N \sim F^{-0.82}$, almost exactly the relationship Wright himself says is predicted by the *plasma hypothesis.*"]

Normally the flux of a source falls off like an inverse square law: $F = A/D^2$, where A is a constant that depends on the *luminosity of the source.* If you count all the sources brighter than a minimum flux F_{min}, then you are looking out to maximum distance $D_{max} = \text{sqrt}(A/F_{min})$. The number of sources varies like D^3, or $N = N_1 (F_{min}/F_1) - 1.5$. *Lerner changes the flux distance relation to* $F = A/D^{2.4}$ *with his added radio absorption,* and this would change the *number count law* to $N = N_1 (F_{min}/F_1) - 1.25$. *If in addition the density of radio sources peaked near the Earth the way that Lerner assumes other densities do,* then the *number count law* becomes $N = N_1 (F_{min}/F_1) - 0.83$. The actual data show $N = N_1 (F_{min}/F_1) - 1.8$ which is *not compatible with Lerner's model.* Thus, *Lerner's local absorbing curtain does not exist.*

Lerner's fit to the *Far Infrared Absolute Spectrophotometer* (FIRAS) spectrum

> [*The Big Bang Never Happened* book was published in 1992 before the referenced *Far Infrared Absolute Spectrophotometer*

(FIRAS) measurements took place. This comment relates to Lerner, E. J. (May, 1995)[5]. *Intergalactic Radio Absorption and the COBE Data*. See above.]

[The objective of the COBE *Far Infrared Absolute Spectrophotometer* (FIRAS) was to measure precisely the *cosmic microwave background* (CMB) spectrum and to observe the *dust* and *line emission* from the *Galaxy*. It was a *polarizing Michelson interferometer* operated differentially with an internal reference *blackbody*, and calibrated by an external *blackbody* having an estimated emissivity of better than 0.9999. It covers the wavelength range from 0.1 to 10 mm in two spectral channels separated at 0.5 mm and has approximately 5% spectral resolution. A flared horn antenna aligned with the COBE spin axis gives the FIRAS a 7-degree field of view. The instrument was cooled to 1.5 K to reduce its thermal emission and enable the use of sensitive bolometric detectors. The *Far Infrared Absolute Spectrophotometer* (FIRAS) ceased to operate when the COBE supply of liquid helium was depleted on September 21, 1990, by which time it had surveyed the sky 1.6 times. The final FIRAS *CMB spectrum* measurements were reported by Fixsen, *et al*. (1996)[6] and Mather, *et al*. (1999)[7] using an improved understanding of the FIRAS calibrator, give the definitive CMB temperature: 2.725 ± 0.002 K.

[6] Fixsen, D. J., *et al*. (December, 1996) The Cosmic Microwave Background Spectrum from the Full COBE FIRAS Data Set. *Astrophys. J*., 473, 576; https://doi.org/10.1086/178173.

[7] Mather, J. C., *et al*. (February, 1999). Calibrator Design for the COBE Far Infrared Absolute Spectrophotometer (FIRAS). *Astrophys. J*., 512, 511-20; https://doi.org/10.1086/306805.]

Assuming the existence of his *absorbing curtain, even though extended distant radio sources show that it does not exist*, Lerner, (1995)[5] presents

185

a fit to the FIRAS spectrum of the *cosmic microwave background* (CMB).

After discussing how there is a slight variation in *"absorbency"* (not defined, units unknown) with *frequency*, Lerner's final fitting function in his Equation (38) assumes an *opacity* that is independent of *frequency*. This function has seven apparent parameters in addition to the 2 parameters of *temperature* and *galactic normalization* that are needed for any FIRAS fit. Lerner then bins the FIRAS data in Mather, *et al.* (1994)[8] from 34 points down to 10 binned points, and finds that his 9-parameter model gives a good fit to 10 data points.

[8] Mather, J. C., *et al.* (1994). Measurement of the cosmic microwave background spectrum by the COBE FIRAS instrument. *Astrophys. J.*, 420, 439-44; https://doi.org/10.1086/173574.

This sounds stupid, but that is mainly due to the paper being poorly written and edited. Lerner's fitting function actually only has two free parameters: a Kompaneets "y" distortion times an *emissivity* that is slightly different from unity. And the resulting 4 parameter fit to the 34 data points in Mather, *et al.* (1994)[8] is pretty good. The Figure below shows the deviation from a blackbody for Lerner's model, and the open circles are the Mather, *et al.* (1994)[8] data.

Unfortunately for Lerner, the improved calibration and use of the full *Far Infrared Absolute Spectrophotometer* (FIRAS) dataset in Fixsen, *et al.* (1996)[5] give the black data points in the Figure. *Lerner's model is a bad fit to this data.* The curve shown, which is the best fit to the Mather, *et al.* (1994)[8] data, is six standard deviations away from the Fixsen, *et al.* (1996)[6] data. Readjusting the emissivity and "y" parameter to best fit the Fixsen, *et al.* (1996)[6] data gives a change in chi[2] of only 0.7 for two new degrees of freedom, which is worse than the average performance of random models.

Lerner's model for the light elements

Lerner wants to make helium in stars. This presents a problem because the stars that actually release helium back into the interstellar medium make a lot of heavier elements too. Observations of *galaxies* with different *helium* abundances show that for every 3.2 grams of *helium* produced, *stars* produce 1 gram of *heavier elements* (French, 1980[9]).

[9] French, H. B. (1980). Galaxies with the Spectra of Giant H II Regions. *Astrophys. J.*, 240, 41-59; https://articles.adsabs.harvard.edu/pdf/1980ApJ...240...41F.

Thus, it is not even possible to make the 28% *helium* fraction in the *Sun* without making four times more than the observed 2% heavier elements fraction, and making the 23% *helium* with only 0.01% of *heavier elements* seen in *old stars* in the *Milky Way halo* is completely out of the question.

[***Lerner's response (below)*** in section headed "Light element production" was "Wright's comments on the *plasma theory* of generating light elements in stars show, again, that *he has not read the relevant papers that he is criticizing*. He assumes that the *distribution of stellar masses* in the early formative periods of galactic history are the same as today, when *supernovae* produce considerable amounts of *carbon–nitrogen–oxygen* (CNO) compared to *helium*. However, the detailed models and calculations presented in my papers showed that the *early galaxies* were dominated by *intermediate mass stars too small to create supernovae*. These *stars* produce and blow off an outer layer of *helium* but very little or no *carbon–nitrogen–oxygen* (CNO) is released to the *interstellar medium*."]

But a further problem is that stars make no lithium and no deuterium. Lerner proposes that these elements are made by *spallation* in *cosmic rays.*

[*Spallation* is a process in which fragments of material (spall) are ejected from a body due to impact or stress. In planetary physics, spallation describes meteoritic impacts on a planetary surface and the effects of stellar winds and *cosmic rays* on planetary atmospheres and surfaces.]

But the *cosmic rays* have 80 *deuterium* nuclei for every *lithium* nucleus (Meyer, 1969)[10] while the Universe has about 6 million *deuterium* nuclei for every *lithium* nucleus.

[10] Meyer, P. (1969). Cosmic Rays in the Galaxy. *Annual Review of Astronomy and Astrophysics*, 7, 1; https://adsabs.harvard.edu/full/1969ARA%26A...7....1M.

So, if the *lithium* is entirely due to *spallation* in *cosmic rays*, the Universe is still missing 99.99% of the observed *deuterium*. Lerner's arithmetic once again fails by a large margin.

[***Lerner's response (below)*** in section headed "Light element production" was "Similar errors occur in Wright's comments on production of *lithium* in *cosmic rays*. Since this occurs when *protons* in *cosmic rays* collide with *carbon–nitrogen–oxygen* (CNO) atoms, naturally the abundance of *lithium* is relatively high in current *cosmic rays*, give the *interstellar medium* contains a few percent CNO. But in very young, formative galaxies, where only one ten-thousandth of the current levels of CNO were yet produced, *Li production was reduced by a comparable amount*. Indeed, we find that stars with heavy element abundance 10^{-4} that of the *sun*, and a few thousand times less than the *interstellar medium* (ISM), have *deuterium/lithium* (D/Li) ratios that are also a few thousand times less than the 80-to-1 ratio Wright quotes. Typically, he misquotes the ratio of *deuterium* (D) to *lithium* (Li) observed in the *oldest stars*, which is about 150,000 to 1, not 6 million to 1."]

Miscellaneous Inconsistencies

- Lerner claims that the *neutrinos* from SN 1987A in the *Large Magellanic Cloud* (LMC) rule out an interesting *neutrino mass*, but the light water detectors used can essentially only detect *electron antineutrinos*, so the *mu* and *tau neutrinos* can have plenty of *mass*. And Lerner's math is wrong (again): A neutrino with an interesting *mass* of 5 eV and an *energy* of 10 MeV travels at a speed that is

$$v = (1 - 0.5*(m/E)^2 + ...)*c = 0.999999999999875*c$$

and after traveling for 160,000 years lags by less than 1 second. The observed burst was 6-10 seconds long, so even the *electron neutrino* could have enough *mass* to be the hot component in a mixed *dark matter* model.
- Lerner shows *cross-sections* of *plasma* simulations that look like a *spiral galaxy* (Figure 1.13 and Figure 6.7). But these are sections of twisting columns as shown in Figure 6.8f, and look nothing like a *spiral galaxy* in 3 dimensions.
- The *Sun* is very massive, and we know the acceleration of the *Sun* as it goes around the *Milky Way*. Therefore, we can compute the force needed to keep it in orbit and compare this to the electromagnetic forces. Thus, it was easy for Ted Bunn to show that *electromagnetic forces* are 100 million billion times too small to affect the orbit of the *Sun* in the *Milky Way*.
- Lerner says about Peebles' calculation of the *helium* abundance that "as the number of *photons* per nucleus increases, so does the production of *helium*." This is just backwards. And the difference between the 30 K predicted by Peebles and the 2.73 K observed now is primarily due to the assumption that the Universe had the critical *density* in *ordinary matter*. So, *this discrepancy is not a failure of the Big Bang, but rather more evidence for non-baryonic dark matter*. And, of course, Peebles did not build the radiometer: Roll and Wilkinson did that.

Lerner, E. J. (July, 2008) Dr. Wright is Wrong-- a reply to Ned Wright's "Errors in *The Big Bang Never Happened*"

https://web.archive.org/web/20160108004949/http://bigbangneverhapp
ened.org/p25.htm.

Retrieved July 13, 2008.

> [The comments in this article relate to the 1992 book "*The Big Bang Never Happened*", not to the online article below which was last modified in 2022.]

A number of people have asked me to reply to Ned Wright's critique[1] of the *Big Bang Never Happened* (BBN).

> [1] Wright, E. L. (October, 2003.) Errors in the "The Big Bang Never Happened". (See above).

Observations since the last edition of the book was published in 1992 have only served to make the arguments in it stronger and to further contradict Wright's assertions.

Large Scale Structures

Wright claims that large scale structures in the universe *can be created in the time since the Big Bang given the existence of dark (non-baryonic) matter in the right amounts.* **There are two errors here. Even calculations by advocates of the *Big Bang* show that *the structures we observe would take about 5 times as long as the Hubble time (the hypothetical time since the Big Bang) to form, even with dark matter.* And, second, *there is no evidence that dark matter exists*.**

Galaxies are organized into *filaments* and *walls* that surround large *voids* that are apparently nearly devoid of all matter. These *voids* typically have diameters around 140-170 Mpc (taking H = 70 km/sec/Mpc) and occur with some regularity[2].

[2] Saar, E., *et al.* (October, 2002). The supercluster-void network V: The regularity periodogram. *Astronomy and Astrophysics*, 393, 1-23; https://doi.org/10.1051/0004-6361:20020811.

These are merely the largest structures commonly observed in present-day surveys of *galaxies*. Still larger structures exist, but are few in number for the simple reason that they are comparable in size with the scope of the surveys themselves.

Since the observed *voids* have galactic *densities* that are 10% or less of the average for the entire observed volume, nearly all the matter would have to be moved out of the *voids*[3].

[3] Hoyle, F., & Vogeley, M. S. (February, 2002). Voids in the Point Source Catalog Survey and the Updated Zwicky Catalog. *Astrophys. J.*, 566, 641-51.

Measurements of the large-scale *bulk streaming velocities* of *galaxies* indicate *average velocities* around 200-250 km/sec[4], a factor of 5 less than the 1,000 km/sec that I conservatively used in my book.

[4] Da Costa, L. N., *et al.* (July, 2000). Redshift-Distance survey of Early-type galaxies: dipole of the velocity field. *Astrophys. J.*, 537, L81-4.

To answer Dr. Wright's objections, let us look at results of large-scale structure formation obtained by his colleagues who support the *Big Bang*, and whose calculations assume that the *Big Bang* happened.

To give the maximum leeway to the *Big Bang* theory, we look at work that assumes some explosive mechanism created the *voids*, which would be much faster than if they were formed by *gravitational attraction*. For a *cold dark matter Big Bang model*, the time T in years, of formation of a *void* R cm in diameter in *matter* with *density* n/cm^3 and final, present-day, *velocity* V cm/s is[5]:

[5] Levin, J. J., Freese, K., & Spergel, D. N. (April, 1992). COBE Limits on Explosive Structure Formation. *Astrophys J.*, 389, 464.

$$T = 1.03 \ n^{-1/4} V^{-1/2} R^{1/2}$$

For V = 220 Km/sec, R = 85 Mpc and n = 2.4 x 10^{-7}/cm^3 (assuming the ratio of baryons to photons, h = 6.14 x 10^{-10}), T= 158 Gy. This is 11.6 times as long as the Hubble time. Even if we increase n to reflect *current assumptions about dark matter* being some 6 times as abundant as ordinary matter, we still get 100 Gy, or 7.4 times the Hubble time. This is actually a bit worse than the figure we arrive at by just dividing the distance moved by the *current velocity*, which ends up as 6.3 time the Hubble time.

Detailed computer simulations, which also include the hypothesized "*cosmological constant*" run into the *same contradictions*, in that *they produce voids that are far too small*. Simulations with a variety of assumptions can produce *voids* as large typically as about 35 Mpc[6], *a factor of 5 smaller than those actually observed on the largest scales*.

[6] Arbabi-Bidgoli, S., & Muller, V. (May, 2002). Void scaling and void profiles in cold dark matter models. MNRAS, 332, 1, 205–214, https://doi.org/10.1046/j.1365-8711.2002.05296.x.

In addition, such simulated voids have *bulk flow velocities* that are typically 10% of the *Hubble flow velocities*[7] which mean that *voids* larger than 60 Mpc, even if they could be produced in *Big Bang* simulations, would generate *final velocities in excess of those observed*, and *voids* as large as 170 Mpc would generate *velocities* of over 600 km/s, *nearly 3 times the observed velocities*.

[7] Schmidt, J. D., Ryden, B. S., & Melott, A. L. (January, 2001). The size and shape of voids in three-dimensional galaxy surveys. *Astrophys. J.*, 546, 2, 609-19.

Thus, even with dark matter AND a cosmological constant, it is impossible for the Big Bang theory to produce voids as large as those observed today with galactic velocities as small as those today. As was true in 1991, *the large-scale structures are too big for the Big Bang.* They in fact *must be far older than the "Big Bang".*

The existence of "dark matter"

Dark matter, or *"non-baryonic" matter* is a *hypothetical form of matter* different from any observed on Earth but *which is nonetheless required by the Big Bang.* Current versions of the (*ever-changing*) theory require that *total gravitating matter density* be equal to 0.3 of the *critical density* but that of ordinary, *baryon matter* be only 0.05 of the *critical density.* This means that 0.25 of the *critical density* has to be in the form of some undiscovered, *non-baryonic matter,* generally described as *Wimps, weakly interacting massive particles.*

This *cold dark matter* (CDM), was hypothesized as essential for the *Big Bang* theory back in 1980--23 years ago. Since then, physicists have searched diligently with dozens of experiments for any evidence of the existence of these dark matter particles here on Earth. Oddly enough *every one of the experiments has had negative results.* In fields of research other than cosmology this would have long ago led to the conclusion that CDM does not exist. But *Big Bang* cosmology does not taken "NO" for an answer. So, the failure to find the CDM after so many experiments does not in any way shake the faith of *Big Bangers* in such *cold dark matter* (CDM). This is evidence that what we are dealing with here is a religious faith, not a scientific theory that can be refuted by experiment or observation.

The idea that *neutrinos* might form a bath of *Hot Dark Matter* has also been undermined by experiments that indicate that while *neutrinos* do probably have some *mass*, it is of the order of 0.1 eV (energy equivalent), which means that *total neutrino mass* in the universe is likely to be around one tenth of the *mass* of ordinary matter.

Wright argues that the existence of *dark matter* is proved by the difference between the *total gravitating mass* inferred for *galaxies* and *cluster of galaxies* and the *mass* in *observable stars*. But this is an absurd non-sequitor. *Observations have demonstrated that stars constitute only a small fraction of the total mass of ordinary matter that can be observed.* In *clusters of galaxies*, we can observe by X-ray emissions *huge clouds of hot plasma*, which have *masses* far greater than that of *bright stars.*

There is extensive observational evidence for *ordinary matter* in two other forms that are relatively dim, one is *white dwarfs* in the *halos* of *spiral galaxies*. Recent observations of *high proper motion stars* have shown that *halo white dwarfs* constitute a *mass* of about 10^{11} *solar masses*, comparable to about half the total estimated mass of the Galaxy[8,9].

[8] Méndez, R. A., & Minnitti, D. (2000). Faint blue objects on the Hubble deep field north and south as possible nearby old halo white dwarfs. *Astrophys. J.*, 529, 911.

[9] Oppenheimer, B. R., *et al.* (March, 2001). Direct Detection of Galactic Halo Dark Matter. *Science*, 292, 698-702.

While these observations have been sharply criticized, they have been confirmed by new observations[10].

[10] Mendez, R. A. (July, 2002). Illuminating the darkness: More evidence from faint-star proper-motions for a cool and warm component to the local dark matter in the Galaxy. arXiv:astrop-ph/0207569.

Observations of *ultraviolet* and *soft x-ray absorption* has revealed the existence of "*warm plasma*" with a temperature of only about 0.2 keV, which amounts to a *mass* comparable to that of the entire *Local group of* galaxies[11].

[11] Nicastro, F., Zezas, A., Elvis, M., *et al.* (2003). The far-ultraviolet signature of the 'missing' baryons in the Local Group of galaxies. *Nature*, 421, 719–721; https://doi.org/10.1038/nature01369.

If we add up the *warm plasma*, which is sufficiently dim to be observable only as it absorbs radiation from more dim objects, the *hot plasma*, and the *white dwarfs, we have enough matter to equal that which is inferred by the gravitational mass of cluster of galaxies. So, there is no need for non-baryonic matter and there is no room for it either.*

Conclusion: *the evidence against the existence of non-baryonic "dark" matter is stronger than ever. Ordinary matter is only the only type of matter that exists.*

A few points on Wright's misunderstanding of the plasma theory of the Cosmic Background Radiation (CBR)

Wright argues that extended radio sources contradict the absorption of radio waves by filaments in the intergalactic medium.* He points to *Cygnus A* and says that no absorbing filaments can be seen. This indicates *Wright has not read the relevant papers*, which make it clear that the *absorbing filaments are quite small by astronomical standards.* Except for an initial 1987 paper, where the idea was worked out only in rough way, my elaboration of the *hypothesis of absorbing magnetic filaments* have made clear that *the filaments in general are too small to be observed directly.* From the formulae in IEEE Transactions on Plasma Science[12], for example, it can be calculated that *filaments* that absorb 21 cm radio waves will be no more than 7,000 km in diameter, *far too small to be resolved.

[12] Lerner, E. J. (December, 1992). Force-Free Magnetic Filaments and the Cosmic Background Radiation. *IEEE Transactions on Plasma Science*, 20, 6, 935-8; https://doi.org/10.1109/27.199554.

Wright's arguing that the inability to resolve the filaments shows their nonexistence is similar to arguing that the inability to resolve individual dust particles in a dust storm contradict the idea that dust absorbs light from the sun.

Wright completely ignores the strong observational evidence that radio emission from galaxies does indeed drop off sharply with distance, relative to emission at IR wavelengths[13], *which are too short to be absorbed by the filaments.*

[13] Lerner, E. J. (September, 1993). Confirmation of radio absorption by the intergalactic medium. *Astrophys. Space Sci.*, 207, 17–26. https://doi.org/10.1007/BF00659126.

He offers no alternative explanation for these observations. This is characteristic of BB theorists, who simply ignore inconvenient observations.

Wright's second objection, *that a fractal inhomogenous collection of absorbers would lead to a non-isotropic distortion of radio sources is simply mathematically wrong.* Fractal distributions are inhomogeneous in three-space, but their projection on to 2-space, the sky, tend to be isotropic.

However, we would expect some fairly small variations in the *Cosmic Background Radiation* (CBR) because of the inhomogenous *Intergalactic Medium* (IGM) -- *where there is more density of matter, we would expect a slightly brighter CBR.* This would only be slight, because scattering and the contribution of the IGM along the same line of sight but at different distance would greatly reduce anisotropies, as described in Lerner, (May, 1995)[14].

[14] Lerner, E. J. (May, 1995). Intergalactic radio absorption and the COBE data. *Astrophys. Space Sci.*, 227, 61-81. https://doi.org/10.1007/BF00678067.

This is what is found. There is indeed a slight correlation between *galaxy density* and *Cosmic Background Radiation* (CBR) intensity, as expected. What is particularly interesting is that *this correlation extends over all angular scales, as would be expected from the plasma viewpoin*t. But in the *Black Body* hypothesis, which assumes the CBR originated BEHIND *all clusters of galaxies* and other *very dense concentrations of matter, interactions with electrons will decrease the CBR luminosity*. So, there should be an anti-correlation of galaxies and CBR on small angular scales. *Just the opposite is observed*[15].

[15] Scranton, *et al.* (July, 2003). Physical Evidence for Dark Energy. arXiv:astrop-ph/0307335.

The correlation continues to be positive even on small angular scales -- as expected in the plasma hypothesis.

In addition, The WMAP results contradict the *Big Bang* theory and support the *plasma cosmology* theory in another extremely important respect. Tegmark, *et al.*[16] have shown that the *quadruple* and *octopole* component of the *Cosmic Background Radiation* (CBR) are not random, but have a strong preferred orientation in the sky.

[16] Tegmark, M., *et al.* (July, 2003). A high resolution foreground cleaned CMB map from WMAP. *Phys. Rev.*, D, 68, 12, 123523; https://doi.org/10.1103/PhysRevD.68.123523.

The *quadruple* and *octopole* power is concentrated on a ring around the sky and are essentially zero along a preferred axis. The direction of this axis is identical with the direction toward the *Virgo cluster* and lies exactly along the axis of the *Local Supercluster filament* of which our Galaxy is a part.

This observation completely contradicts the Big Bang assumption that the Cosmic Background Radiation (CBR) *originated far from the local Supercluster and is, on the largest scale, isotropic without a preferred direction in space. Big Bang* theorists have implausibly labeled the coincidence of the preferred CBR direction and the direction to Virgo to be mere accident and have scrambled to produce new ad-hoc assumptions, including that the universe is finite only in one spatial direction, an assumption that *entirely contradicts the assumptions of the inflationary model of the Big Bang*, the only model generally accepted by *Big Bang* supporters.

However, *the plasma explanation is far simpler*. If the *density* of the *absorbing filaments* follows the *overall density of matter*, as assumed by this theory, then the *degree of absorption* should be higher *locally* in the direction along the axis of the (roughly cylindrical) *Local Supercluster* and lower at right angles to this axis, where less *high-density matter* is encountered. This in turn means that concentrations of the *filaments outside the Local Supercluster*, which slightly enhances *Cosmic Background Radiation* (CBR) power, will be more obscured in the direction along the supercluster axis and less obscured at right angle to this axis, as observed. More work will be needed to estimate the magnitude of this effect, but *it is in qualitative agreement with the new observations*.

Wright's third objection illustrates the essential sloppiness of *Big Bang* thinking. He claims that statistics of flux vs number counts contradict the *absorption hypothesis*. *In fact they confirm it*. Contrary to Wright's claims that $N \sim F^{-1.8}$, where N is the number of sources brighter than F, *the actual distribution is quite different*. Wright's formula is roughly true ONLY for the very brightest sources, those stronger than about 200 mJy. *For sources dimmer than that*, the relationship is very close to $N \sim F^{-0.82}$, almost exactly the relationship Wright himself says is predicted by the *plasma hypothesis*[17].

[17] Windhorst, R., *et al.* (March,1993). Microjansky source counts and spectral indices at 8.44 GHz. *Astrophys. J.*, 405, 2, 498-517.

Wright either is ignorant of this well-known fact, or deliberately ignores it.

There is no real mystery as to why the brighter sources follow a different relationship. As Sylos Labini, *et al.*[18] demonstrate, for very bright sources, the *number-flux relationship* is distorted by finite size effects.

[18] Labini, S., *et al.* (May, 1996). Finite size effects on the galaxy number counts: Evidence for fractal behavior up to the deepest scale. *Physica*, A, 226, 195-242.

Put simply, *very bright sources are either very close or, if distant and intrinsically bright, are very rare.* For small volumes there will be too few of these very bright objects -- for small enough volumes there will be none of them. As the volume increases to the size at which a fair sample of very bright objects occurs, the apparent density increases. This creates a purely apparent, statistical tendency for a more rapid growth in the number of objects with decreasing flux. The true relationship is only revealed with the more numerous dimmer objects.

A very similar change in the *number flux slope* occurs in the counts of *optical sources*, basically *galaxies*, with one important different. For *bright galaxies*, the relationship has an exponent of -1.5, but for *dim galaxies*, the exponent changes to -1.0. That exponent is just what one would expect for a fractal distribution of dimension $D = 2$ with NO *absorption*. The fact that the radio sources have an exponent of -0.82, not -1.0, implies an *absorption* almost identical to that hypothesized in the *plasma theory of the Cosmic Background Radiation* (CBR) *power. Without absorption, one would have to explain why more distant radio sources become systematically dimmer and less numerous compared with optical course* -- even at distances of tens of Mpc, far too small to be affected by evolutionary effects.

A Brief Note on the Hubble relationship

Wright says that my book endorses Alfven's explanation of the Hubble relationship. *But again, that implies that Wright did not even read the book he criticizes.* In the book, I present Alfven's, AND several other explanations of the Hubble relationship in the Appendix to the book (which was in both editions), as well as in Chapter 6. I concluded that "the question of the Hubble relationship remains unanswered" (p. 279) and that none of the possible explanations were without problems, a conclusion that still stands. However, *the one explanation that can be ruled out, because of its many contradictions with observation, is the Big Bang.* We are not stuck with the *Big Bang* by default.

Light element production

In considering the arguments against the Big Bang (BB), Wright entirely ignores the contradictions between observations and Big Bang (BB) predictions of light element abundances, pointed out in the preface to my book. *These contradictions have only gotten sharper since the book was written. ...*

Big Bang Nucleosynthesis (BBN) predicts the abundance of four light isotopes (^{4}He, ^{3}He, D and ^{7}Li) given only the *density of baryons* in the universe. *These predictions are central to the theory, since they flow from the hypothesis that the universe went through a period of high temperature and density -- the Big Bang. In practice, the baryon density has been treated as a free variable, adjusted to match the observed abundances.* Since four abundances must be matched with only a single free variable, the light element abundances are a clear-cut test of the theory. *In 1992, there was no value for the baryon density that could give an acceptable agreement with observed abundances, and this situation has only worsened in the ensuing decade.* The current observations of just three of the four predicted BBN light elements preclude BBN at a level of at least 7 σ. In other words, *the odds against BBN being a correct theory are about 100 billion to one.*

Wright's comments on the *plasma theory* of generating light elements in stars show, again, that *he has not read the relevant papers that he is criticizing*. He assumes that the *distribution of stellar masses* in the early formative periods of galactic history are the same as today, when *supernovae* produce considerable amounts of *carbon–nitrogen–oxygen* (CNO) compared to *helium*. However, the detailed models and calculations presented in my papers showed that the *early galaxies* were dominated by *intermediate mass stars too small to create supernovae*. These *stars* produce and blow off an outer layer of *helium* but very little or no *carbon–nitrogen–oxygen* (CNO) is released to the *interstellar medium*[19].

[19] Lerner, E. J. (April, 1989). Galactic Model of Element Formation. *IEEE Transactions on Plasma Science*, 17, 3, 259-63; https://doi.org/10.1109/27.24633.

Similar errors occur in Wright's comments on production of *lithium* in *cosmic rays*. Since this occurs when *protons* in *cosmic rays* collide with *carbon–nitrogen–oxygen* (CNO) atoms, naturally the abundance of *lithium* is relatively high in current *cosmic rays*, give the *interstellar medium* contains a few percent CNO. But in very young, formative galaxies, where only one ten-thousandth of the current levels of CNO were yet produced, *Li production was reduced by a comparable amount*. Indeed, we find that stars with heavy element abundance 10^{-4} that of the *sun*, and a few thousand times less than the *interstellar medium* (ISM), have *deuterium/lithium* (D/Li) ratios that are also a few thousand times less than the 80-to-1 ratio Wright quotes. Typically, he misquotes the ratio of *deuterium* (D) to *lithium* (Li) observed in the *oldest stars*, which is about 150,000 to 1, not 6 million to 1. But to a true *Big Bang* believer like Dr. Wright, making an error of a factor of forty in regards to mere observations is no cause of concern. Observations, after all, do not affect faith.

Fixsen, D. J. (December, 2009). The Temperature of the Cosmic Microwave Background.

Astrophys. J., 707, 2, 916–920; https://doi.org/10.1088/0004-637X/707/2/916.

University of Maryland, Goddard Space Flight Center, MD, USA.

Received October 6, 2009.
Accepted October 28, 2009.

Abstract

The *Far InfraRed Absolute Spectrophotometer* (FIRAS) data are independently recalibrated using the *Wilkinson Microwave Anisotropy Probe data* to obtain a *cosmic microwave background* (CMB) temperature of 2.7260 ± 0.0013. Measurements of the *temperature* of the CMB are reviewed. The determination from the measurements from the literature is CMB temperature of 2.72548 ± 0.00057 K.

---

1. INTRODUCTION

The *Wilkinson Microwave Anisotropy Probe* (WMAP) data present an opportunity to recalibrate the *Far-InfraRed Absolute Spectrophotometer* (FIRAS) experiment and produce an independent check of the other measurements of the *cosmic microwave background* (CMB) *temperature*. In Sections 2 and 3, the WMAP data will be presented and combined with the FIRAS data to make an independent estimate of the CMB temperature. In Sections 4 and 5, this new estimate will be combined with others from the literature to generate an improved estimate for the CMB temperature.

The WMAP data only measure the difference in intensity between different points on the sky. However, the precision is sufficient such that

202

the *velocity* of the WMAP spacecraft can be used to calibrate the *velocity* to various points on the *surface of last scattering* of the *cosmic microwave background* (CMB).

> [The reference to the *surface of last scattering* can be ignored. *According to the Big Bang theory*, the *surface of last scattering* is a spherical shell-shaped region around us from which the photons of the *cosmic microwave background* (CMB) are arriving. It is the shell-shaped region of space at the distance that the *photons* necessarily came from. The *surface of last scattering* corresponds to a virtual outer surface of the spherical observable universe. It refers to a shell at the right distance in space so *photons* are now received that were originally emitted at the time of *decoupling*. *According to the Big Bang theory*, the *time of decoupling* is a period in the development of the universe when different types of particles fall out of thermal equilibrium with each other. This occurs as a result of the *expansion of the universe*, as their interaction rates decrease (and mean free paths increase) up to this critical point. The two verified instances of decoupling since the *Big Bang* which are most often discussed are *photon decoupling* and *neutrino decoupling*, as these led to the cosmic microwave background and cosmic neutrino background, respectively.]

This velocity in turn is used to form a differential spectrum of the CMB. The differential spectrum is then fit with a single parameter which is the *cosmic microwave background* (CMB) *temperature*.

2. WMAP VELOCITY MAP

The standard *Wilkinson Microwave Anisotropy Probe* (WMAP) sky maps (Hinshaw, *et al.* 2009[1]) are corrected to the baricenter of the *solar system* using the *Jet Propulsion Laboratory* (JPL) ephemeris (Standish & Fienga, 2002[2]).

[1] Hinshaw, G., *et al*. (February, 2009, Five-Year Wilkinson Microwave Anisotropy Probe Observations: Data Processing, Sky Maps, and Basic Results. *Astrophys. J. Supp.*, 180, 2, 225-45; https://doi.org/10.1088/0067-0049/180/2/225.

[2] Standish, E. M., & Fienga, A. (March, 2002). Accuracy limit of modern ephemerides imposed by the uncertainties in asteroid masses. *Astronomy and Astrophysics*, 384, 1, 322-8; https://doi.org/10.1051/0004-6361:20011821.

The calibration assumes a *cosmic microwave background* (CMB) temperature of 2.725 K (derived from the *Far InfraRed Absolute Spectrophotometer* (FIRAS) measurement). However, the various changes as the *Wilkinson Microwave Anisotropy Probe* (WMAP) makes its way around the Sun (now in its ninth repetition) can be used to calibrate the WMAP data in terms of velocity. The velocity of the WMAP spacecraft, with respect to the Sun, is known to < 1 cm s^{-1}, which is a negligible uncertainty compared to other uncertainties considered here. This velocity is used to calibrate a map of velocity relative to the *surface of last scattering* of the CMB. Most of this velocity is the *dipole*, presumably *the motion of the solar system with respect to the frame of the CMB*; however, it includes temperature variations due to the Sachs–Wolfe effect which has the same spectrum. This process is repeated for each WMAP differential assembly and each year yielding 50 independent maps for the first five years of WMAP operation.

The WMAP *velocity* maps are the *temperature* maps, available at http://lambda.gsfc.nasa.gov/product/map/current/m_products.cfm, with the dipole added back in and divided by the CMB temperature. In generating the WMAP *temperature* maps, a *dipole* is fit and removed from the raw data. The residual variations are due to various sensitivities of the WMAP instrument—the changing velocity as the spacecraft makes its annual trek around the solar system and the small variations of the CMB as a function of position. …

…

4. CMB TEMPERATURE

There were many publications of measurements of the *cosmic microwave background* (CMB) temperature from the late '60s and '70s, but the uncertainties are large and the systematics were not well understood. Here an arbitrary cutoff of 50m K uncertainty was used to select 15 sources for the CMB temperature from recent publications. These results are shown in Table 2.

...

5. SUMMARY AND CONCLUSIONS

The calibration methods for the FIRAS have been described and the accuracy estimated. All of the recent precision estimates of the *cosmic microwave background* (CMB) temperature agree within 2.5 times their uncertain ties. These estimates were made with a variety of methods from different platforms and different frequencies. Combining all of the estimates, results in a very modestly elevated χ^2 and an improved *absolute temperature estimation* of 2.72548 ± 0.00057 K.

Ali-Hamoud, Y.* (November 12, 2012[#]). Spinning dust radiation: a review of the theory.

https://arxiv.org/pdf/1211.2748.

\# Later version at https://doi.org/10.1155/2013/462697.

* Institute for Advanced Study Princeton, New Jersey 08540, USA

The author is supported by the National Science Foundation grant number AST-080744 and the Frank and Peggy Taplin Membership at the Institute for Advanced Study.

Abstract

This article reviews the current status of theoretical modeling of *electric dipole radiation* from *spinning dust grains*. The fundamentally simple problem of *dust grain rotation* appeals to a rich set of concepts of classical and quantum physics, owing to the diversity of processes involved. *Rotational excitation* and *damping rates* through various mechanisms are discussed, as well as methods of computing the *grain angular momentum distribution function*. Assumptions on grain properties are reviewed. The robustness of theoretical predictions now seems mostly limited by the uncertainties regarding the *grains* themselves, namely their abundance, dipole moments, size and shape distribution.

---

1 Introduction

Rotational radiation from small grains in the *interstellar medium* (ISM) has been suggested as a source of *radio emission* several decades ago already. The basic idea was first introduced by Erickson (1957)[1] and then revisited by Hoyle & Wickramasinghe (1970)[2] and Ferrara & Dettmar (1994)[3].

206

[1] Erickson, W. C. (November, 1957). A Mechanism of Non-Thermal Radio-Noise Origin. *Astrophys. J.*, 126, 480. See above.

[2] Hoyle, F., & Wickramasinghe, N. C. (August, 1970). Radio Waves from Grains in H_{II} Regions. *Nature*, 227, 473-4.

[3] A. Ferrara, A., & Dettmar, R.-J. (May, 1994). Radio-emitting dust in the free electron layer of spiral galaxies: Testing the disk/halo interface. *Astrophys. J.*, 427, 155-9. See abstract above.

Rouan, *et al.* (1992)[4] were the first to provide a thorough description of the physics of rotation of *polycyclic aromatic hydrocarbons* (PAHs), although not including all gas processes.

[4] Rouan, D., Leger, A., Omont, A., & M. Giard. (January, 1992). Physics of the rotation of a PAH molecule in interstellar environments. *Astronomy and Astrophysics*, 253, 498-514.

Shortly after the discovery of the *anomalous dust-correlated microwave emission* (AME) in the galaxy by Leitch et al (1997)[5], Draine and Lazarian (1998, hereafter DL98[6,7]) suggested that spinning dust radiation might be responsible for the AME, and provided an in-depth theoretical description of the process.

[5] Leitch, E. M., Readhead, A. C. S., Pearson, T. J., & Myers, S. T. (September, 1997). An Anomalous Component of Galactic Emission. *Astrophys. J.*, 486, L23-26; https://www.aoc.nrao.edu/~smyers/papers/1997ApJL.Leitch.486.L23-26.pdf.

[6] Draine, B. T., & Lazarian, A. (February, 1998). Diffuse Galactic Emission from Spinning Dust Grains. *Astrophys. J.*, 494, L19–L22; https://iopscience.iop.org/article/10.1086/311167/pdf. See above.

[7] Draine, B. T., & Lazarian, A. (November, 1998). Electric Dipole Radiation from Spinning Dust Grains. *Astrophys. J.*, 508, 157-79; https://doi.org/10.1086/306387.

Understanding the spinning dust spectrum in as much detail as possible is important. First, the *anomalous microwave emission* (AME) constitutes a *foreground emission* to *cosmic microwave background* (CMB) radiation. Second, it provides a window into the properties of small grains, which play crucial roles for the physics and chemistry of the *interstellar medium* (ISM).

Motivated by these considerations and the accumulating observational evidence for diffuse and localized AME, several groups have since then revisited and refined the DL98 model (...). New physical processes were accounted for, that can significantly affect the predicted spectrum. ... .

The purpose of this article is to provide an overview of the physics involved in modeling spinning dust spectra. We attempt to provide a comprehensive description of the problem at the formal level, and let the interested reader learn about the details in the various works that deal with the subject.

This article is organized as follows: Section 2 reviews the basic process of *electric dipole radiation* and the resulting *emissivity*. We then describe the assumed properties of the *small grains* which are believed to be the source of the spinning dust radiation in Section 3. Section 4 discusses the *rotational configuration* of small grains stochastically heated by ultraviolet (UV) photons. Section 5 describes the methods to obtain the distribution of *grain angular momentum*, as well as the various physical processes that affect it. We conclude and mention potential future research directions in Section 6.

...

6 Concluding remarks

In this article we have reviewed the current status of *spinning dust modeling*, and tried to summarize the recent advances in this field since the seminal papers of Draine & Lazarian[6,7]. In addition to refined calculations, the most important new effect accounted for recently is grain wobbling following frequent absorption of UV *photons*. The rotational dynamics of *small grains* of various shapes is now believed to be well understood, even if there remain uncertainties and simplifications in the implemented models.

The accuracy of theoretical predictions remains mostly limited by our poor knowledge of the properties of small grains, namely their dipole moments, shapes and sizes, as well as their overall abundance, about which other observations give little information. This uncertainty can be turned into an asset, as one could potentially use the observed spinning dust emission (*assuming it is the dominant anomalous microwave emission* (AME) *process* at tens of GHz frequencies) to constrain properties of *small grains*.

Such a procedure can, however, only be accomplished if environmental parameters are very well known. Indeed, the *gas density, temperature* and *ionization state* as well as the *ambient radiation field* all affect the *rotational distribution function* of *small grains* in non-trivial ways. In addition, the actual observable, the *emissivity*, depends upon the properties of the medium along the line of sight, and an accurate modeling of the spatial properties of the environment is also required. Unless the properties of the environment are well understood, it seems very difficult to extract dust grain parameters from observed spectra, due to the important degeneracies that are bound to be present for such a large parameter space.

The view of the author is that significant advances in the field would be possible if several regions of the *interstellar medium* (ISM) were put under the scrutiny, not only of radio telescopes, but also of instruments at other wavelengths, in order to determine their detailed properties as much as possible and get rid of the uncertainties related to environmental dependencies.

Finally, let us mention another potentially interesting avenue to probe the properties of emitting grains, namely the high-resolution *spectral properties* of the spinning dust spectrum. Indeed, even if the *polycyclic aromatic hydrocarbons* (PAHs) are classical rotators with large rotational quantum numbers, the line spacing remains relatively large for the smallest molecules (for coronene for example, rotational lines are spaced by about 0.33 GHz). A large number of different grains are probably present in the *interstellar medium* (ISM), which results in a dense, quasi-smooth forest of lines. However, grains with a few tens of atoms might only be present in a limited number of stable configurations, or there might only be a fraction of possible grain configurations that lead to a significant *electric dipole moment*. If this were the case, radio observations with a narrow bandwidth should allow us to detect some amount of bumpiness on top of a smooth spectrum. Even upper limits on the variability of the spectrum in the frequency domain should allow one to get some handle on the properties of *small grains*. A quantitative analysis of this issue will be the subject of future work.

Acknowledgements

I thank Bruce Draine and Alexander Lazarian for providing detailed comments on this manuscript, as well as Rashid Sunyaev for his hospitality and generous financial support at the Max Planck Institute for Astrophysics during part of summer 2012, where and when this article was written.

Kroupa, P.*, Pawlowski, M.*, & Milgrom, M.[#] (December, 2012). The failures of the standard model of cosmology require a new paradigm.

Int. J. Mod. Phys., D, 21, 1230003; https://arxiv.org/pdf/1301.3907.

* Argelander-Institute for Astronomy, University of Bonn, Auf dem Hugel 71, Bonn,
# Department of Particle Physics Germany.and Astrophysics Weizmann Institute of Science Rehovot 76100, Israel.

Received October 22, 2012.
Revised November 18, 2012.

———————————

Abstract

Cosmological models that invoke warm or cold dark matter cannot explain observed regularities in the properties of dwarf galaxies, their highly anisotropic spatial distributions, nor the correlation between observed mass discrepancies and acceleration. These problems with the standard model of cosmology have deep implications, in particular in combination with the observation that the data are excellently described by *Modified Newtonian Dynamics* (MOND). MOND is a classical dynamics theory *which explains the mass discrepancies in galactic systems, and in the universe at large*, without invoking 'dark' entities. MOND introduces a new universal constant of nature with the dimensions of acceleration, a_0, such that the pre-MONDian dynamics is valid for accelerations $a \gg a_0$, and the deep MONDian regime is obtained for $a \ll a_0$, where space-time scale in *variance* is invoked. Remaining challenges for MOND are (i) *explaining fully the observed mass discrepancies in galaxy clusters*, and (ii) *the development of a relativistic theory of MOND that will satisfactorily account for cosmology*. [Why? ???] The *universal constant a_0* turns out to have an intriguing connection with cosmology: $a_0^- \equiv 2\pi a_0 \approx cH_0 \approx c^2(\Lambda/3)^{1/2}$.

211

This may point to a deep connection between cosmology and internal dynamics of *local systems*.

---

1. Introduction

Newton[1] formulated his dynamics subject to the observational constraints from the Solar system, while Einstein developed his *general relativistic field equation*[2] with the prerequisite that they reproduce Newton's work in the classical, non-relativistic limit.

[1] Newton, I. (1760). *Philosophiae naturalis principia mathematica*, vol. 1- 4.
[2] Einstein, A. (1916). Die Grundlage der allgemeinen Relativitätstheorie. *Ann. Phys.*, 354, 769.

General relativity has successfully passed tests *in the astronomically small spatial-scale and strong acceleration limit* [???] (Solar system and smaller in length scale and stronger in acceleration scale).

The presently favored understanding of cosmological physics is based on assuming Einstein's field equation to hold on galactic and larger scales and for very small accelerations as are found in galactic systems (postulate 1). In addition, *it is assumed that all matter emerged in the Big Bang* (postulate 2).

The first two postulates lead to an inhomogeneous and highly curved cosmological model which is in disagreement with the observed distribution of matter, unless inflation is additionally postulated to drive the universe to near flatness and homogeneity briefly after the Big Bang (postulate 3). The *structures* and their *kinematics* which arise from this postulate again *do not match the observed ones unless cold or warm dark matter (DM) is postulated in addition to aid gravitational clumping on galactic scales* (postulate 4). This *dark matter* (DM) must, *as observations constrain, consist of non-relativistic particles which only interact significantly through gravitation.*[3]

212

3 Blumenthal, G. R., Faber, S. M., Primack, J. R., & Rees, M. J. (October, 1984). Formation of galaxies and large-scale structure with cold dark matter. *Nature*, 311, 517-25; 525 (1984). https://doi.org/10.1038/311517a0.

The so constructed model *does not match the observed increase in the rate of expansion* [???] unless it is additionally postulated that *dark energy* drives an emerging new era of *inflation* (postulate 5).

The resulting five-postulate construction with its many parameters determined using observational data[4] is referred to as the *concordance cosmological model* which is the currently generally accepted *cold or warm DM-based standard model of cosmology* (SMoC).[5,6]

4 Famaey, B., & McGaugh, S. S. (September, 2012). Modified Newtonian Dynamics (MOND): Observational Phenomenology and Relativistic Extensions. *Living Rev. Relativ.*, 15, 10; https://doi.org/10.12942/lrr-2012-10.

5 Frenk, C. S., & White, S. D. M. (October, 2012). Dark matter and cosmic structure. *Ann. Phys.*, 524, 507-34; https://doi.org/10.1002/andp.201200212.

6 J. Silk, J., & Mamon, G. A. (August, 2012). The Current Status of Galaxy Formation. *Research in Astronomy and Astrophysics*, 12, 8, 917-46; https://doi.org/10.1088/1674-4527/12/8/004.

The SMoC can be used to compute the *distribution of matter on galactic scales* because it is based on *Newtonian dynamics* which is a linear and thus, in principle, readily computable dynamics theory.[5,7]

7 Kuhlen, M., Vogelsberger, M., & Angulo, R. (October, 2012). Numerical Simulations of the Dark Universe: State of the Art and the Next Decade. arXiv:astro-ph/1209.5745; https://doi.org/10.48550/arXiv.1209.5745.

The complex processes of the *baryons* (*heating, cooling, star-formation, stellar and other feedback*) are dealt with by employing parametrized laws, *but few rigorous convincing predictions have come forth* due to the tunability of the modelling, the lack of adopted constraints from observed *star formation processes*, and due to the haphazardness of the many mergers each *dark matter* (DM) *halo* experiences.[8]

[8] Mutch, S. J., Poole, G. B., & Croton, D. J. (January, 2013). Constraining the last 7 billion years of galaxy evolution: In semi-analytic models. *MNRAS*, 428, 3, 2001–16; https://doi.org/10.1093/mnras/sts182.

The *statistical distribution of sub-halos on the scales of many kpc around normal galaxies* is among the rigorous predictions of the SMoC. The other rigorous prediction is that *dwarf galaxies formed from tidal material expelled during galaxy encounters* (see Sect. 2) *cannot contain substantial amounts of dark matter* (DM).

The computations by many research groups have demonstrated that the SMoC has significant problems, the number of which has been increasing with improving computer power.[9]

[9] Kroupa, P. (June, 2012). The Dark Matter Crisis: Falsification of the Current Standard Model of Cosmology. *PASA*, 29, 395-433; https://doi.org/10.1071/AS12005.

While the greatest problems (see below) are usually not discussed by dark matter (DM) *advocates even when they do discuss problems for the SMoC*, e.g. in Ref (5), it is held by most *dark-matter* advocates that virtually all problems mentioned by the respective authors *can be solved once modelling the baryonic processes is improved with larger resolution.*[7]

Thus, for example, one of the failures, the *missing satellite problem*, is deemed to have been solved by now: according to the SMoC each *dark matter* (DM) *halo* ought to contain a large number of *satellite DM halos* which are the *phase-space substructures* that merge during the hierarchical formation of larger structures.[13]

[13] Klypin, A. A., Trujillo-Gomez, S., & Primack, J. (October, 2011). Dark matter halos in the standard cosmological model: Results from the bolshoi simulation. *Astrophys. J.*, 740, 102; https://doi.org/ 10.1088/ 0004-637X/740/2/102.

Milky Way and Andromeda class galaxies ought to have many hundreds of satellite galaxies, each of which are immersed in their own *dark matter* (DM) *halos*. The observed small number of satellites (currently 24 are known for the *Milky Way* and a slightly larger number is known for *Andromeda*) is then a result of the complex and tunable *baryonic processes* cleaning out *baryons* from the vast majority of *satellite* DM *halos*.

Some of the problems of the SMoC have been used to investigate possible model extensions, for example by invoking new dark forces between the *dark matter* (DM) particles which affect structure growth on sub-kpc scales. These additional postulates do not significantly affect the distribution of *sub-halos* on scales of many kpc, as this is given by the hierarchical infall of DM into larger structures.

The SMoC has been stated to be in agreement with the *large-scale distribution of matter (although the problems on the local 100 Mpc and 16 Mpc scales undermine this statement*, Sec. 4) and with the *cosmic microwave background* (CMB) *fluctuations*.[5] But it can only remain as a valid description of the real universe if every one of the postulates individually pass observational tests. In order to certify the validity of the SMoC and its dark-sector variants robustly against the observational data, tests must be developed which are insensitive to the details of *baryonic physics*. Various such tests have recently been devised,[15] and here the two most important ones (Sec. 2, Sec. 3) with consistency checks (Sec. 4) are discussed.

[15] Kroupa, P., *et al.* (November, 2010). Local-Group tests of dark-matter concordance cosmology. Towards a new paradigm for structure formation. *Astronomy and Astrophysics*, 523, A32, 22; https://doi.org/ 10.1051/0004-6361/201014892.

The observed correlation of kinematical properties of *galaxies* over several orders of magnitude in acceleration (Sec. 5) convincingly demonstrates that a *gravitational theory* which differs from Newton's is valid. *The Modified Newtonian Dynamics (MOND) theory*[16] *is in excellent agreement with these data and provides an apparently correct description of dynamics on galaxy scales* (Sec 6).

[16] Milgrom, M. (July, 1983). A modification of the Newtonian dynamics as a possible alternative to the hidden mass hypothesis. *Astrophys. J.*, 270, 365-70; https://doi.org/10.1086/161130.

2. The dual dwarf galaxy theorem

In any realistic cosmological theory two types of *galaxies* emerge. *Primordial ('type A') galaxies* form *directly after the Big Bang* [???] from *gravitational instabilities* in the cooling *baryonic matter*. *Tidal dwarf galaxies ('type B')* form from interacting, rotationally-supported *type A galaxies*: from tidal arms, which fragment and form *dwarf galaxies* as well as *star clusters*. This is evident in many observations of interacting *galaxies* and in all high-resolution simulations of interacting *galaxies*.

Indeed, high-resolution simulations in the SMoC show that *type A* and *type B dwarfs* have different dynamical and morphological properties (see points (i) and (ii) below). This comes about because the formation of *type A dwarfs* is dominated by the collapse and mergers of dwarf *dark matter* (DM) *halos* in which the end products retain much of the DM. In the SMoC, a large number of these *sub-halos* orbit as bound type A sub-structures in the more massive host *halos*, many more than are observed: The *missing satellite problem* therewith emerges (Sec. 1). *Type B dwarfs*, on the other hand, cannot capture significant amounts of DM because they form from the 'cold' parts of (disc) *galaxies*, which exclude the *dark matter* (DM). In the SMoC so many *tidal dwarf galaxies* are generated that to account for these, statistically, most of the observed *dwarf elliptical* (dE) *galaxies* would have to be *type B dwarfs*.

The SMoC thus predicts that two types of *galaxies* ought to be abundant that have different dynamics: one showing *dark matter* (DM), the other not. But in fact of observation, rotating late-type *dwarf galaxies* as well as faint *dwarf-spheroidal galaxies* show evidence for *dark matter* (DM). A *tidal dwarf galaxy* which is observed to have a DM component does not immediately disprove the SMoC. *Dark matter* (DM) may appear to be present in a *tidal dwarf galaxy* if it is wrongly assumed to be in dynamical equilibrium when, in fact, it is not, as it is being perturbed by a time varying *tidal field* from the host *galaxy*. And, a young *tidal dwarf galaxy* may be accreting *gas* and may thus not be in rotational equilibrium feigning a DM content.

All *type A galaxies* which are rotationally supported are known to precisely lie on the *Baryonic-Tully-Fisher Relation* (BTFR)[4] which therefore must be defined by the DM halo *if the SMoC is true. The physics responsible for the precise conspiracy of the putative DM halo and galaxy luminosity to produce the BTFR over many orders of magnitude in galaxy luminosity however remains unknown. Tidal dwarf galaxies* that show evidence for *dark matter* (DM) because they are out of equilibrium ought not to lie on the BTFR, because the *tidal perturbation* of, or *gas accretion* onto, the *dwarf* would not likely conspire to place a DM-free *type B galaxy* onto the BTFR of the *type A dwarfs*.

The prediction made by the SMoC that there are two distinct classes of dwarf galaxy (the 'dual dwarf galaxy' theorem) is falsified by observations, as follows:[9]

(i) By invoking the *Baryonic-Tully-Fisher Relation* (BTFR)[4] of *rotationally supported galaxies*: *Type B dwarfs* do not contain *dark matter* (DM) and therefore cannot lie on the same BTFR of *type A galaxies* in the SMoC. *But observations show them to lie on the same BTFR as type A galaxies. The rotational velocities* of the *type A galaxies* therefore cannot be given by a DM halo. Therewith, *galactic DM halos cannot exist* and one of the major pillars of the SMoC (postulate 4) breaks away *such that the whole model is falsified.*

(ii) By considering the *radius–mass relation* of *pressure-supported dwarf galaxies*: *dwarf elliptical* (dE) *galaxies* are thought to be of type A and thus to form in putative *dark matter* (DM) halos. *The final structural parameters of dE galaxies and type B dwarfs therefore cannot agree. But observations show type B dwarfs to be indistinguishable from dE galaxies.* Thus, *DM halos cannot play a role in dwarf galaxies*, leading to the same conclusion as under (i) above.

3. Significant anisotropic phase-space distribution of satellite galaxies

The SMoC robustly predicts *the primordial dark matter* (DM) *dominated (type A) satellite galaxies in a major halo to be distributed spheroidally around the host galaxy.* The infall of *satellite galaxies* from *filaments* leads only to mild anisotropies. The *Milky Way galaxy*, being part of the universe, must conform to these predictions. Instead, the Galactic satellite system forms a *vast polar structure* (VPOS) which is a disk-like distribution about 40 kpc thick and 400 kpc in diameter containing all known *satellite galaxies* as well as all *young halo globular clusters* and about half of all known *stellar and gas streams.*[30]

[30] Pawlowski, M. S., Pflamm-Altenburg, J., & Kroupa, P. (June, 2012). The VPOS: a vast polar structure of satellite galaxies, globular clusters and streams around the Milky Way. *MNRAS*, 423, 2, 1109-26; https://doi.org/10.1111/j.1365-2966.2012.20937.x.

The *proper motion* measurements of 10 *satellite galaxies* show 9 of the *satellites* to be orbiting within the *vast polar structure* (VPOS) (one *satellite* is on a *counter-rotating orbit* in comparison to the 8 others). One satellite, *Sagittarius*, orbits perpendicularly to the VPOS and to the *Milky Way* (MW). *To have nine of ten type A (DM sub-halo) satellites with measured proper motions orbiting within the VPOS is ruled out with very high confidence.*[31]

[31] Libeskind, N. I., Frenk, C. S., Cole, S., Jenkins, A., & Helly, J. C. (October, 2009). How common is the Milky Way-satellite system alignment? *MNRAS*, 399, 550-8; https://doi.org/10.1111/j.1365-2966.2009.15315.x.

Andromeda also has an *anisotropic satellite* system, and a few other *galaxies* are known which have *satellite galaxies* arranged in highly correlated phase-space structures.[9] *The prediction of the SMoC is therewith ruled out conclusively because the Milky Way is not unique in this property.*[9,15,34]

[34] Angus, G. W., Diaferio, A., & Kroupa, P. (September, 2011). Using dwarf satellite proper motions to determine their origin. *MNRAS*, 416, 14010-9; https://doi.org/10.1111/j.1365-2966.2011.19138.x

4. Consistency tests

The falsification[9] of the SMoC with the above argument has deep implications. Can this argument be erroneous? A number of auxiliary tests have been developed,[15] *which the SMoC again does not pass.* A visualization of the results of the comparison between many predictions and observational data is presented in the theory confidence graph.[9] *If each of the 22 listed failures were to be associated with a loss of confidence of (only) 50 per cent that the SMoC accounts for a particular property of the real universe, then according to the confidence graph, the overall probability that the SMoC is a valid theory of the universe would be 0.5 raised to the power of 22, corresponding to a confidence of 2.3×10^{-5} per cent.*

Three examples of such problems are the *cusp/core problem, the downsizing* and *the missing-bright-satellites problem.* These are usually acknowledged as problems, but claimed to be (probably) solvable.[5,7] *The real problems for which no solution can be identified are usually not mentioned in the dark matter* (DM) *literature*: (a) The *dual dwarf galaxy theorem* and *satellite anisotropy* failures discussed above *are catastrophic for the* SMoC. (b) In the Local Volume with radius of

8 Mpc around the Sun, *the matter distribution in the Local Void and near the edges of the filaments containing galaxies is incompatible with the expectations from the* SMoC.37 (c) *The 50 Mpc radius region around us is missing DM by a factor of 3–4 while the fluctuations in the density must not be larger than about 10 per cent on these scales if the* SMoC *were true.* (d) *Disk galaxies of similar mass are observed to show too little scatter* in their properties, which *implies that galaxies obey laws that do not emerge in the* SMoC.

5. The mass-discrepancy-acceleration correlation: towards galaxy laws

The *rotation velocity*, V(r), measured at a *radius* r in a *disk galaxy* can be compared to the *rotation velocity*, V_b(r), that one would expect due to only the *observed baryonic matter*. A plot of V^2/V_b^2, which is the *mass discrepancy* at r, in a given *galaxy* as a function of r, gives the putative *dark matter* (DM) profile of the galaxy. As pointed out by McGaugh,[40] combining many such measurements for many galaxies yields a mass discrepancy vs r plot with no correlation.

[40] McGaugh, A. S. (July, 2004). The Mass Discrepancy-Acceleration Relation: Disk Mass and the Dark Matter Distribution. *Astrophys. J.*, 609, 2, 652-66; https://doi.org/10.1086/421338.

However, plotting the *mass-discrepancy* in dependence of the *baryonic-Newtonian acceleration*, $g_N = V_b^2/r$, yields a tight correlation (Fig. 1). The *mass-discrepancy–acceleration* correlation deviates from the Newtonian value below an acceleration of $a_0 \approx 1.2 \times 10^{-10}$ m/s² = 3.8 pc/Myr² which is about five orders of magnitude smaller than the acceleration of *Neptune* in the *Solar system*.

No known physics of the dark matter (DM) *particles can account for the observed mass discrepancy–acceleration correlation.* But this relation is convincing evidence for *gravitational dynamics* becoming non-Newtonian at accelerations $a < a_0$. This and many other correlations were predicted 30 years ago by Milgrom[16] who introduced a new constant of nature, a_0 at the foundation of a new paradigm dubbed '*Modified Newtonian Dynamics*' (MOND). *In this paradigm Newtonian dynamics and general relativity are modified, so that galaxy dynamics is explained without the need for dark matter.* Since its formulation in 1983[16] MOND has passed all observational tests from 10^6 M$_\odot$ *dwarf disk galaxies*[4] to *massive elliptical galaxy*[41] scales, with some tension remaining on the *globular- and galaxy-cluster scales*.[9]

[41] Milgrom, M. (September, 2012). Testing MOND over a Wide Acceleration Range in X-Ray Ellipticals. *Phys. Rev. Lett.*, 109, 131101; https://doi.org/10.1103/PhysRevLett.109.1311.

These are, however, not major.[42,43]

[42] Sanders, R. H. (September, 2007). Neutrinos as cluster dark matter. *MNRAS*, 380, 1, 331-8; https://doi.org/10.1111/j.1365-2966.2007.12073.x
[43] Sanders, R. H. (May, 2012). NGC 2419 does not challenge MOND, Part 2. *MNRAS*, 422, 1, L21-23; https://doi.org/10.1111/j.1745-3933.2012.01227.x.

6. The MOND paradigm

While MOND has developed significantly in its particulars (such as formulations of various underlying theories), its basic *non-relativistic* (NR) tenets, from which follow all its major predictions, remain the same as originally proposed. These tenets, put in a somewhat improved form, are: (1) A new constant, a_0, with the dimensions of *acceleration* is introduced into dynamics. (2) A MOND theory must tend to standard dynamics in the limit $a_0 \to 0$ (in other words, when the theory is applied to a system where all the quantities of dimensions of *acceleration*, a, are much larger than a_0). (3) In the opposite limit, $a_0 \to \infty$ (namely, when all $a \ll a_0$ in which case we also have to take G $\to$ 0 so that Ga_0 remains fixed) the theory is postulated to become *space-time scale invariant*,[44] namely, invariant under $(t,\mathbf{r}) \to \lambda(t,\mathbf{r})$.

[44] Milgrom, M. (June, 2009). The Mond Limit from Spacetime Scale Invariance. *Astrophys. J.*, 698, 2, 1630-8; https://doi.org/10.1088/0004-637X/698/2/1630.

Conceptually, a_0 thus plays a role as a demarcation acceleration between the validity domain of the pre-MOND dynamics ($a \gg a_0$) and the MOND regime ($a \lesssim a_0$). In the former, a_0 disappears from physics, but in the latter domain a_0 appears with full impact on various phenomena. *These roles are similar to those of Planck's constant in the classical–quantum context, and of the speed of light, c, in the classical–relativistic context.* All these constants play the role of demarcations between the *old* and *new physics*. Also, taking them to the appropriate limit ($\hbar \to 0$ – the *correspondence principle*, and c $\to \infty$) takes one to the old, *classical theory*, in which these constants do not appear. Also, they all appear ubiquitously in the description of many, apparently unrelated phenomena[a] in the *new-physics* regime.

[a] For example, *in quantum theory*: the *black-body spectrum*, the *photoelectric effect*, the *hydrogen spectrum*, *quantum Hall effect*, etc.

We shall see below how a_0 enters in many such phenomena and effects in *low acceleration galactic dynamics*.

Newtonian gravitational accelerations, which scale as $g_N \propto MG/r^2$, transform under the above *space-time scaling* as $g_N \to \lambda^{-2} g_N$, while *kinematic accelerations* ($\ddot{x}$) transform as g $\to \lambda^{-1}$g. *Newtonian dynamics*, which equates the two, is thus *not scale invariant*. To have such a symmetry in the MOND limit we need g to scale like $g_N^{1/2}$; or, with the help of a_0, g $\propto (a_0 g_N)^{1/2}$. This, more primitive, description of the MOND limit, is essentially the original formulation of MOND for test-particle motion.[16] The formulation of the 3rd tenet in terms of scale invariance,[44] is, however, rather more elegant, precise, and general, and should be preferred.

Beyond the above basic tenets, one wishes to construct a full-fledged theory, *and then extend it to the relativistic regime.* [Why?] Several *relativistic* and *non-relativistic* (NR) MOND theories are known. In the *non-relativistic* (NR) regime, we have the suitably chosen nonlinear extension of the *Poisson equation*,[45] and a quasilinear MOND formulation (QUMOND).[46]

[45] Bekenstein, J., & Milgrom, M. (November, 1984). Does the missing mass problem signal the breakdown of Newtonian gravity? *Astrophys. J.*, 286, 7-14; https://doi.org/10.1086/162570.

[46] Milgrom, M. (April, 2010). Quasi-linear formulation of MOND. *MNRAS*, 403, 2, 886-95; https://doi.org/10.1111/j.1365-2966.2009.16184.x.

These are classified as '*modified gravity*' (MG) theories as they modify *the field equations of the gravitational field*, but not *the laws of motion*. There are also '*modified inertia*' (MI) formulations, which do the opposite. For the latter there isn't yet a full-fledged theory, but, nonetheless, much can be said about their predictions of rotation curves. *In the relativistic regime* we can mention as examples, TeVeS, and its predecessors, MOND adaptations of *Einstein aether theories*, bimetric MOND (BIMOND), one based on a polarizable medium, and non-local, single metric formulations.

Much work has also been done over the years towards devising *observational tests* of MOND and on comparing MOND predictions with observations. Extensive recent reviews of these aspects of the MOND paradigm exist.[4,54]

[54] Sanders, R. H., & McGaugh, S. S. (2002). Modified Newtonian Dynamics as an Alternative to Dark Matter. *ARAA*, 40, 263-317; https://doi.org/10.1146/annurev.astro.40.060401.093923.

Quantum theory and relativity were not mere changes in form of the equations of classical dynamics. They each brought about totally new concepts to underlie dynamics. Likewise, there are reasons to presume that what we know about MOND today is only the tip of an iceberg.[44] One hint that this might be so is the '*coincidence*'[16] $a_0^- \equiv 2\pi a_0 \approx cH0 \approx c^2(\Lambda/3)^{1/2}$, where H_0 is the *Hubble constant* and Λ is the *cosmological*

223

constant (CC). This, and some aspects of symmetry, may point to a deep connection between dynamics within *local systems*, such as *galactic systems*, and the state of the universe at large. This connection, if firmly established, could be the most far-reaching implication of MOND.[b]

[b] Such a cosmological connection might imply that various aspects of the MOND-standard dynamics interplay may depend on cosmic time.

The MOND theories mentioned above depart from pre-MOND dynamics in that they introduce a_0, add new degrees of freedom, and modify the underlying action, but they do not deeply depart in spirit from their predecessor. Perhaps one of them will turn out to be an effective theory that captures the essence of the future deeper MOND theory.[c]

[c] The appearance in all these theories of an interpolating function is an indication that they can only be effective theories.

…

7. MOND vs dark matter

We have seen that the SMoC faces many difficulties in accounting for various aspects of data on galaxies (Sec. 2– 4). But perhaps the most cogent argument against the SMoC is that MOND works[62,64,65] (e.g. Fig. 1).

[62] Milgrom, M. (March, 2008). The MOND paradigm. ArXiv:astro-ph/0801.3133; https://doi.org/10.48550/arXiv.0801.3133.
[64] Milgrom, M. (May, 2008). Marriage à-la-MOND: Baryonic dark matter in galaxy clusters and the cooling flow puzzle. *New Astronomy Reviews*, 51, 10-12, 906-15; https://doi.org/10.1016/j.newar. 2008.03.023.
[65] Sanders, R. H. (November, 2009). Modified Newtonian Dynamics: A Falsification of Cold Dark Matter. *Advances in Astronomy*, https://doi.org/10.1155/2009/752439.

This is not because MOND is a competitor, but because the many tight regularities and correlations between *baryon* and '*dark matter* (DM)' properties, predicted by MOND, and confirmed in the data, are quite contrary to what is expected in the *dark matter* (DM) paradigm. In the DM paradigm, a present-day system is the haphazard end result of a complex history that affects *baryons* and *dark matter* (DM) very differently. The observed tight correlations argue against this.[39]

It is important to emphasize that *in MOND, the dynamics of a self-gravitating system is strictly predicted once its baryon distribution is given.* But in the SMoC it is not possible to predict the dynamics of an individual *primordial (type A) galaxy* just from its observed *baryons*; at best one can try to match a *dark matter* (DM) *halo* from an infinite family that will fit the observed dynamics. For example, given the *baryon distribution* of a *disc galaxy*, the MOND rotation curve is a prediction, while the DM curve is but a fit with much freedom allowed. In short, unlike for MOND, the SMoC cannot predict the relations between *baryons* and *dark matter* (DM) in any individual system because these depend on the unknowable history of the system. In the rare case where the SMoC does make a prediction (*tidal dwarfs*, i.e. *type B galaxies*), because their formation histories are well understood, *it fails.*

Angus[67] and others have demonstrated that *MOND-based cosmological models are able to account for many observations such as the CMB and the Bullet cluster.*

[67] Angus, G. W., & Diaferio, A. (October, 2011). The abundance of galaxy clusters in modified Newtonian dynamics: cosmological simulations with massive neutrinos. *MNRAS*, 417, 941-9; https://doi.org/10.1111/j.1365-2966.2011.19321.x.

8. MOND and the dwarfs

In MOND the *dual dwarf galaxy theorem* predicts no systematic dynamical differences between *primordial* and *tidal dwarf galaxies*. For example, both rotational types A and B of dwarfs are predicted to fall on the same *Baryonic-Tully-Fisher Relation* (BTFR) by MOND law 2 in

Sect. 6, irrespective of their different formation paths. MOND is fully consistent with the existence of the *vast polar structure* (VPOS) around the Milky Way (Sec. 3) as being an *ancient tidal structure* in which the Galaxy's *satellites* formed as *tidal dwarf galaxies* and *globular clusters*. *Tidal dwarf galaxies* form naturally in MONDian galaxy interactions. *The internal constitution of the Galaxy's dwarf-spheroidal satellites is well explained with MOND.* No such consistent explanation can be achieved within the SMoC, because the high measured Newtonian *mass-to-light ratios* of the *satellites* are completely in contradiction with them being ancient *tidal dwarf galaxies*. This, however, is the only physically plausible explanation for the existence of the *vast polar structure* (VPOS) with counter-orbiting satellites.[72]

[72] Pawlowski, M. S., Kroupa, P., & de Boer, K. S. (May, 2011). Making counter-orbiting tidal debris. The origin of the Milky Way disc of satellites? *Astronomy and Astrophysics*, 532, A118; https://doi.org/10.1051/0004-6361/201015021.

9. Conclusions

The SMoC is based on five postulates (Sec. 1); it is reported as having appreciable success in describing the *large-scale structure* and the *cosmic microwave background* (CMB). More importantly, it has a long history of *grave failures*[f].

[f] Compare to the *standard model of particle physics* which is known to be incomplete but since decades it has consistently been found to be in excellent agreement with the data.

The failures are not really surprising since the first and most fundamental postulate is equivalent to an extrapolation of the empirically established gravitational law by Newton and Einstein to scales many orders of magnitude below the Solar system acceleration scale. Instead, the data correlations unequivocally show dynamics at accelerations smaller than $a_0 \approx 1.2 \times 10^{-10}$ m/s^2 to not be Einsteinian/Newtonian. Given the nature of the failures, *dynamically significant cold or warm DM particles cannot be the explanation of the mass discrepancies*. Observations strongly suggest that a complete theory of dynamics has to depart from standard dynamics below some critical acceleration that is a constant of nature. The extraordinary success of MOND in accounting for the observational data on *galactic scales* and its properties (Sec. 6–8) suggest that the expected underlying theory will contain a deep connection between the dynamics within *local systems* and the state of the universe at large. Understanding the deeper physical meaning of MOND remains a challenging aim. It involves the realistic likelihood that *a major new insight into gravitation will emerge, which would have significant implications for our understanding of space, time and matter.*

Planck's Cosmic Microwave Background (CMB) Map

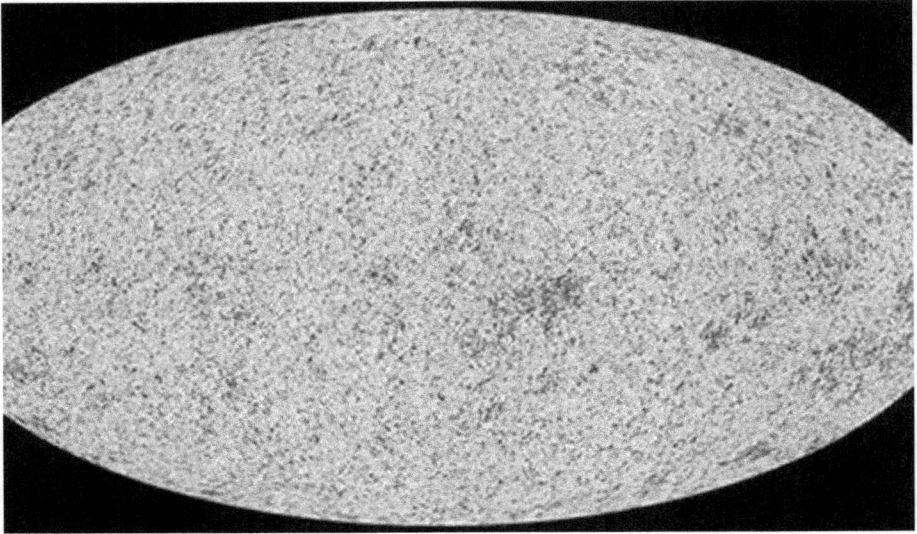

This map shows small temperature fluctuations of the *cosmic microwave background* (CMB) radiation that correspond to regions of slightly different densities, representing the seeds of the stars and galaxies of today. This map was constructed from an analysis of observations of the sky at wavelengths of light spanning 850 microns to 1 cm (353 GHz to 30 GHz). Additional observations spanning 350 to 550 microns (857 to 545 GHz) helped characterize foreground dust in the Milky Way, which was removed from the final CMB data shown here.

This view of the data is in an equirectangular projection suitable for projection onto a sphere, and is useful for full-dome presentations. The projection is in galactic coordinates with the galactic plane running horizontally along the midpoint of the image. Most graphics software will map images to the outside of a sphere; this is the inside projection looking outwards.

Image courtesy of ESA/NASA. Planck is a European Space Agency (ESA) mission, with significant participation from NASA.

Schwarz D. J.[1], Copi, C. J.[2], Huterer, D.[3], & Starkman, G. D.[2] (August, 2016). CMB anomalies after Planck.

Class. Quantum Grav., 33, 184001; https://doi.org/10.1088/0264-9381/33/18/184001; also at https://websites.umich.edu/~huterer/Papers/CQG_anomalies.pdf.

[1] Fakultät für Physik, Universität Bielefeld, Postfach 100131, D-33501 Bielefeld, Germany.

[2] CERCA/Department of Physics/ISO, Case Western Reserve University, Cleveland, OH 44106-7079, USA.

[3] Department of Physics, University of Michigan, 450 Church St, Ann Arbor, MI 48109-1040, USA.

DJS is supported by the DFG grant RTG 1620 'Models of Gravity'. GDS and CJC are supported by the US Department of Energy grant DOE-SC0009946. DH is supported by NSF under contract AST-0807564 and DOE under contract DE-FG02-95ER40899.

Received October 21, 2015.
Revised April 4, 2016.
Accepted for publication June 20, 2016.

Abstract

Several unexpected features have been observed in the microwave sky at large angular scales, both by WMAP and by Planck. Among those features is a *lack of both variance and correlation on the largest angular scales, alignment of the lowest multipole moments with one another and with the motion and geometry of the solar system*, a *hemispherical power asymmetry or dipolar power modulation*, a *preference for odd parity modes* and an unexpectedly *large cold spot in the Southern hemisphere*.

The individual p-values of the significance of these features are in the per mille to per cent level, when compared to the expectations of the

229

best-fit inflationary [*relativistic*] ΛCDM model. Some pairs of those features are demonstrably uncorrelated, increasing their combined statistical significance and indicating *a significant detection of CMB features at angular scales larger than a few degrees on top of the standard model*. Despite numerous detailed investigations, we still lack a clear understanding of these large-scale features, which seem to imply a *violation of statistical isotropy and scale invariance of inflationary perturbations*. In this contribution we present a critical analysis of our current understanding and discuss several ideas of how to make further progress.

[*Cosmic variance* is the *statistical uncertainty* inherent in *observations* of the universe at extreme distances. The most widespread use reflects the fact that *measurements are affected by cosmic large-scale structure*, so a measurement of any region of sky (viewed from Earth) may differ from a measurement of a different region of sky (also viewed from Earth) by an amount that may be much greater than the sample variance. This is based on the idea that *it is only possible to observe part of the universe at one particular time*, so it is difficult to make statistical statements about cosmology on the scale of the entire universe, as the number of observations (sample size) must be not too small. It is sometimes used to mean *the uncertainty because we can only observe one realization of all the possible observable universes*. For example, we can only observe one *Cosmic Microwave Background*, so the measured positions of the peaks in the *Cosmic Microwave Background* spectrum, integrated over the visible sky, are limited by the fact that *only one spectrum is observable from Earth*. The observable universe viewed from another galaxy will have the peaks in slightly different places, while remaining consistent with the same physical laws, inflation, etc.

The *lack of correlation* refers to the fact that the Planck observations showed *no correlation with simulations based on*

230

the ΛCDM model for angular scales $70° < \theta < 170°$ and a significant lack of correlation at $\theta > 60°$.

The *alignment of the lowest multipole moments with one another and with the motion and geometry of the solar system* refers to the fact that the quadrupole and octopole are found to be mutually aligned and they define axes that are unusually perpendicular to the Ecliptic pole and parallel to the direction of our motion with respect to the rest frame of the *cosmic microwave background* (CMB) (the *dipole* direction).

The *hemispherical power asymmetry or dipolar power modulation* refers to the observation that the power in discs on the sky of radius $10°$ - $20°$ evaluated in several multipole bins, is larger in one hemisphere on the sky than the other. The plane that maximizes the asymmetry is approximately the Ecliptic, though it depends somewhat on the multipole range. At all multipole moments (ℓ's), *there appears to be more power in the north sky than in the south sky*. Sometimes it is assumed that the power may be modulated along the sky by a *dipole* term.

Preference for odd parity modes refers whether the CMB sky is symmetric to reflections around the origin. The standard theory does not predict any particular behavior with respect to this *point-parity symmetry*, but determines that even (odd) *multipoles* have an even (odd) *symmetry*. Observations indicate a preference for *odd parity* modes is found in some features, such as a *lack of variance and correlation on the largest angular scales, alignment of the lowest multipole moments*, and a *hemispherical power asymmetry or dipolar power modulation*.

The reference to an *unexpectedly large cold spot in the Southern hemisphere* relates to a cold spot with a radius of about five degrees centered on *angular coordinates* $(l, b) = (207°, -57°)$, for which the evidence for its existence is frequency independent. The size of the cold spot makes it too large to be a point source, yet typically too small to be a diffuse foreground, especially since it is found in a relatively foreground-clean part of the sky. And while the Sunyaev-Zeldovich effect inverse Compton scattering of the *cosmic microwave background* (CMB) photons off hot electrons in galaxy clusters could in principle lead to the desired amplitude and spatial extent of the signal, the Sunyaev-Zeldovich effect has a very pronounced frequency dependence that is completely incompatible with the observed frequency independence. However, it is possible that the cross-correlation performed more selectively e.g. looking for CMB over-density behind clusters of galaxies or voids, or behind the cold spot alone, would show departures from ΛCDM predictions. Such tests performed to date have shown tantalizing, though as yet not definitive, evidence for a large under-density in the distribution of galaxies in the same direction as the cold spot.]

---

1. Introduction

Among the purposes of this contribution is to summarize the evidence for *unexpected features of the microwave sky at large angular scales*, as revealed by the observation of temperature anisotropies by the space missions *Cosmic Background Explorer* (COBE), *Wilkinson Microwave Anisotropy Probe* (WMAP) and *Planck*. Before doing so, let us put those discoveries into context with the study of other aspects of the *cosmic microwave background* (CMB) radiation.

[This article is focused on anomalies in the CMB *background radiation*, rather than *foreground anomalous microwave emission* (AME) discussed above.]

Half a century ago, the discovery of the *cosmic microwave background* (CMB) revealed that most of the *photons* in the Universe belong to a highly *isotropic thermal radiation* at a temperature of ~3K. Deviations from this *isotropy* were first found in the form of a *temperature dipole* at the level of ~3 mK. This *dipole* has been interpreted as the effect of

232

Doppler shift and aberration due to the *proper motion* of the *solar system* with respect to a *cosmological rest frame*.

The observation of an isotropic *cosmic microwave background* (CMB), together with the *proper-motion hypothesis*, provides strong support for the *cosmological principle*. This states that the Universe is *statistically isotropic and homogeneous*, and restricts our attention to the *Friedmann–Lemaître* class of [*relativistic*] cosmological models [???]. The *cosmological principle* itself is a logical consequence of the *observed isotropy* and the Copernican principle, the statement that we are typical observers and thus observers in other galaxies should also see a nearly isotropic CMB.

The *proper-motion hypothesis* is supported by the (*Cosmic Background Explorer*) COBE *satellite* discovery of *higher multipole moments*.

> [The *proper motion hypothesis* refers to the astrometric measure of the observed changes in the apparent places of stars or other celestial objects in the sky, as seen from the *center of mass* of the *Solar System*, compared to the abstract background of the more distant stars.
>
> A *multipole expansion* is a mathematical series representing a function that depends on angles—usually the two angles used in the spherical coordinate system (the polar and azimuthal angles) for three-dimensional Euclidean space. Similarly to Taylor series, *multipole expansions* are useful because oftentimes only the first few terms are needed to provide a good approximation of the original function. The function being expanded may be real- or complex-valued.
>
> *Multipole expansions* are used frequently in the study of electromagnetic and gravitational fields, where the fields at distant points are given in terms of sources in a small region. The *multipole expansion* with *angles* is often combined with an

expansion in *radius*. Such a combination gives an expansion describing a function throughout three-dimensional space.

The *multipole expansion* is expressed as a sum of terms with progressively finer angular features (*moments*). The first (the zeroth-order) term is called the *monopole moment*, the second (the first-order) term is called the *dipole moment*, the third (the second-order) the *quadrupole moment*, the fourth (third-order) term is called the *octupole moment*, and so on.]

These higher *moments* turned out to be two orders of magnitude below the *dipole* signal, at a rms *temperature fluctuation* of ~30 μK at 10° *angular resolution*. However, a direct test of the *proper-motion hypothesis* had to wait until *Planck* was able to resolve the *Doppler shift* and aberration of hot and cold spots at the smallest angular scales It is important to note here that the *observed dipole* could also receive contributions from effects other than the solar system's *proper motion*. These could be as large as 40% without contradicting the *Planck* measurement at the highest *multipole moments*. Observations at *non-cosmic microwave background* (CMB) frequencies, e.g. in the *radio* or *infrared*, hint at significant *structure dipoles* or *bulk flows*, but are still inconclusive. Here we dwell on this aspect as the CMB *dipole* is one of the most important calibrators in modern cosmology. It defines what we call the CMB *frame* and many cosmological observations and tests refer to it.

The existence of *structures* like *galaxies*, *voids* and *clusters* imply that the *cosmic microwave background* (CMB) cannot be perfectly *isotropic*. The COBE discovery [COBE satellite discovery of *higher multipole moments*] revealed the long-expected *temperature anisotropies* and confirmed that they are consistent with an almost *scale-invariant power spectrum* of *temperature* fluctuations. *Scale invariance* of the temperature anisotropies means that the band *power spectrum* $D_t = \ell\,(\ell + 1)\,C_\ell/2\pi$ is a constant for small *multipole number* ℓ. Here C_ℓ denotes the expected *variance* in the *amplitude* of any spherical harmonic component of the *temperature fluctuations* with *total angular-momentum** ℓ.

> * This analogy from quantum physics is useful to describe the spherical harmonic analysis of temperature fluctuations in terms of well-known physical concepts.

During the last two decades, ground-based, balloon-borne and satellite *cosmic microwave background* (CMB) experiments led to an improved understanding of those *temperature* anisotropies. The WMAP and *Planck* space missions played a special role, obtaining full-sky measurements that enabled us to investigate a large range of *angular scales*, from the *dipole* ℓ = 1 to ℓ ~ 2500, more than three decades in ℓ. The band power spectrum as published by Planck is shown in figure 1.

[*According to the Big Bang theory*,] these *temperature fluctuations* are believed to have been generated from quantum fluctuations in the very early Universe by a (nearly) scale-invariant mechanism. The most prominent context is *cosmological inflation*. If *inflation* lasts long enough, the spatial geometry of the Universe is generally predicted to be indistinguishable from Euclidean, and the topology of the observable Universe is expected to be trivial (simply connected). Even more importantly, *inflation* predicts that the CMB *temperature fluctuations* should be: (i) *statistically isotropic*, (ii) *Gaussian*, and (iii) almost *scale invariant*. It also predicts: (iv) *phase coherence* of the fluctuations; (v) for the simplest models, *a dominance of the so-called adiabatic mode* (strictly speaking it is not only adiabatic but also isentropic); and (vi) the *non-existence of rotational modes at large scales*. Finally, depending on the energy scale of *cosmological inflation*, there might be (vii) a detectable stochastic background of *gravitational waves* that also obeys properties (i)–(iii).

In the process of extracting cosmological parameters from the *cosmic microwave background* (CMB) and other observations, properties (i)–(vi) are assumed to hold true and a stochastic *gravitational wave background* is neglected. [*According to the Big Bang theory*,] this leads to the minimal inflationary [*relativistic*] Λ cold dark matter (ΛCDM) model.

Analysis of the *cosmic microwave background* (CMB) allows us not only to fit all free parameters of this [*relativistic*] model, but also to *test its underlying assumptions. However, the more fundamental the assumption, the harder it appears to test*. The existence of the peaks and dips shown in figure 1 are due to the *phase coherence*, property (iv). The almost-scale-invariance (iii) is visible in the smallness of the deviations from the best-fit model, although *a model-independent reconstruction of the primordial power spectrum leaves room for deviations at the largest observed scales*. More detailed analysis also reveals that there is a *strong upper limit of at most 4% of non-adiabatic modes* (v), while *rotational modes would have produced a large B-polarization signal that is not observed*. The predicted flatness and the expected trivial topology are consistent with all observations.

It thus remains to test *Gaussianity* and *statistical isotropy*. A lot of effort has been put into searches for *non-Gaussianity* and they are described in great detail elsewhere. The brief summary is that *there is no evidence for it so far. In the following we focus our attention on statistical isotropy, and touch on the issue of scale invariance*.

All mentioned predictions should hold at all observable scales. However, [*according to the Big Bang theory*,] testing these primordial properties of the Universe directly *is complicated by physics related to the evolution of the Universe after the end of cosmological inflation*. In order to understand which phenomena can be most cleanly probed at which scales it is instructive to look at the comoving size corresponding to a particular angular scale as a function of *redshift*, see figure 2.

[*According to the Big Bang theory*,] at the time of the formation of the first atoms, scales that today subtend more than about a degree (and that therefore affect $\ell \leq 180$) were not much affected by details of *photon decoupling*. Thereafter, the Universe was filled with a mix of H and He gas, until it was *reionized* at a *redshift* of about 10. *Angular scales* larger than about 20° (or $\ell \leq 10$) are also not much affected by the details of *reionization*. Finally, *angular scales* larger than $\sim 60°$ (or $\ell \leq 3$) enter the *Hubble scale* at a *redshift* of one and thus are either of *primordial* or *local* origin. Here by *local* we mean from within our *Hubble patch* of the Universe. Thus, it is a good idea to start with a test of *statistical isotropy* at the largest *angular scales*, as whatever we find must be either *primordial* or a *local* effect due to either *foreground* or *local cosmic structure*.

In this contribution we intend to give a summary of the evidence for the existence of features of the microwave sky that apparently violate statistical isotropy on the largest angular scales (section 2). Since this seems to happen only at the largest *angular scales*, it also amounts to a *violation of scale invariance*. We also discuss several ideas that have been put forward to explain those features, though we do not intend to give an exhaustive review. Apart from the suggestion that all of them are statistical flukes (the probability for which to happen is tiny, unless compensated for by huge look-elsewhere penalties) these ideas can be classified into *foreground effects* (section 3) and *cosmological effects* (section 4). In section 5 we highlight several possible tests of those ideas. The study of *polarization* at large *angular scales* and more detailed all-sky study of *non-CMB wavebands* seems to be particularly promising.

2. A summary of the evidence

...

2.1. Low variance and lack of correlation

Historically, the first surprise, already within the COBE data, was the smallness of the *quadrupole moment*. When WMAP released its data, it confirmed C_2 to be low, however it was also shown that *cosmic variance** allows for such a small value.

> * *Cosmic variance* is defined for $\ell \geq 2$. For $\ell = 0$ it diverges. For $\ell = 1$ it would be well defined, but *cosmic variance* does not

apply if the CMB *dipole* is caused by the *proper-motion* of the solar system.

Another rediscovery in the first release of WMAP was that the angular two-point correlation function at *angular scales* $\geq 60°$ is unexpectedly close to zero, *where a non-zero correlation signal was to be expected.* This feature had already been observed by COBE, but was ignored by most of the community before its rediscovery by WMAP. ...

...

2.2. Alignments of low multipole moments

In the standard [*relativistic*] ΛCDM model the *temperature* (and other) anisotropies have random phases. In harmonic space this means that *the orientations and shapes of the multipole moments are uncorrelated.* This was first explored in the first year WMAP data release using the *angular momentum dispersion* where it was discovered that the *octopole* ($\ell = 3$) is somewhat planar (dominated by m = ± ℓ for an appropriate choice of coordinate frame orientation) with a p-value of about 5%. (A *quadrupole* is always *planar.*) More importantly, the normal to this plane (the axis around which the *angular momentum dispersion* is maximized) was found to be surprisingly *well aligned with the normal to the quadrupole plane* at a p-value of about 1.5%.

...

The most recent analysis of the latest WMAP and the Planck 2013 data releases finds *the quadrupole and octopole anomalously aligned with one another*, with p-values ranging from about 0.2%–2% depending on the exact map employed. It is further found that *the quadrupole and octopole are jointly perpendicular to the ecliptic plane* (i.e. their area vectors are *nearly orthogonal to the normal to the ecliptic*) with a p-value of 2%–4% and *to the Galactic pole* with a p-value of 0.8%–1.6%. Even more strikingly *they are aligned with the dipole direction* with a p-value of 0.09%–0.37%.

...

In summary, *the octopole is unexpectedly planar; the quadrupole and octopole planes are unexpectedly aligned with each other, and unexpectedly perpendicular to the ecliptic and aligned with the CMB dipole*. These *alignments* have been robust *in all full-sky data sets* since WMAP's first release, and are found to be exacerbated by proper removal of the *kinetic quadrupole*. *No systematics and no foregrounds have been identified to explain these apparent violations of statistic isotropy.*

2.3. Hemispherical asymmetry

Evidence for *hemispherical power asymmetry* first emerged in the analysis of WMAP first year data. It was found that *the power in disks on the sky of radius ~10°–20°, evaluated in several multipole bins, is larger in one hemisphere on the sky than the other*; see the left panel of figure 5. *The plane that maximizes the asymmetry is approximately the ecliptic*, though it depends somewhat on the *multipole* range; the variation of the normal to this plane with multipole range is shown in the right panel of figure 5. Figure 4 shows that the combined *quadrupole* and *octopole moment* already contribute to such a *power asymmetry*.
…

The fact that the axis that maximizes the *asymmetry* is close to the *ecliptic pole* motivates both systematic and cosmological proposals for the *hemispherical asymmetry*. Nevertheless, *there have been no convincing proposals to date about why one ecliptic hemisphere should have less power than the other.*

2.4. Parity asymmetry

It is interesting to ask *whether the CMB sky is symmetric with respect to reflections around the origin* … . The *standard theory* does not predict any particular behavior with respect to this *point-parity symmetry*. Because $Y_{\ell m}(-\hat{e}) = (-1)^{\ell} Y_{\ell m}(\hat{e})$, even (odd) *multipoles* ℓ have an even (odd) *symmetry*.
…

Using a suitably defined *power spectrum* statistic—the ratio of the sum over multipoles of D_ℓ for the even map to that for the odd map—they found *a 99.7% evidence for the violation of parity* in WMAP7 data in the *multipole range* $2 \leq \ell \leq 22$. The analysis was finally extended to *Planck* … , who confirmed the results … based on WMAP, but also found that the significance depends on the *maximum multipole* chosen, and peaks for $\ell_{max} \simeq 20\text{--}30$, but is lower for other values of the *maximum multipole* used in the analysis; see figure 7. …

…

Whether the observed parity asymmetry is a fluke, an independent anomaly, or a byproduct of another anomaly, is not clear at this time. The *parity asymmetry* appears to be correlated with the *missing power at large angular scales*, as the wiggles in the lowest *multipoles*, seen clearly in the top panel of figure 7, combine to nearly perfectly cancel the *angular two-point correlation function* above $60°$.

2.5. Special regions: the cold spot

Evidence for an unusually *cold spot* in WMAP 1st year data was first presented in [73].

[73] Vielva, P., Martinez-Gonzalez, E., Barreiro, R. B., Sanz, J. L., & Cayon, L. (July, 2004). Detection of Non-Gaussianity in the Wilkinson Microwave Anisotropy Probe First-Year Data Using Spherical Wavelets. *Astrophys. J.*, 609, 1, 22–34; https://doi.org/10.1086/421007.

The spot, shown in figure 8, is centered on *angular coordinates* $(l, b) = (207°, -57°)$, has a *radius* of approximately five degrees, is roughly circular, and the evidence for its existence is *frequency independent*. The *cold spot* was originally detected using spherical Mexican hat wavelets, which are well suited for searching for compact features on the sky; tests … detected a deviation from the *Gaussian expectation* in the kurtosis of the wavelet coefficients at the wavelet scale of $R = 300^l$. Taking into account the 'look elsewhere' effect, that is the fact that not all statistics attempted with the wavelets returned an anomalous result, [75] estimated the statistical level of anomaly of the *cold spot* to be 1.85%.

[75] Cruz, M., Cayon, L., Martinez-Gonzalez, E., Vielva, P., & Jin, J. (2007). The non-Gaussian cold spot in the 3 year Wilkinson Microwave Anisotropy Probe data. *Astrophys. J.*, 655, 1, 11–20; https://doi.org/ 10.1086/509703.

...

If the cold spot is indeed taken as a sign of departure from the [relativistic] ΛCDM model's predictions, it may be possible to explain it using novel theory. ...

...

3. Foregrounds

If these anomalous *cosmic microwave background* (CMB) features are related to *local* physics, it might not be surprising that they appear to be a rare fluke. The reason is simple—our environment is one particular example of an environment for a CMB mission and every particular realization is somewhat special. In this section we review some of the *local* physical effects that have been suggested as explanations for CMB *anomalies*. We ignore speculations on instrumental effects, as it seems to us that the consistency of WMAP and *Planck* 2015 results makes such an explanation quite unlikely.

3.1. Solar system

The closest *foreground* to a *cosmic microwave background* (CMB) space mission is the solar system itself. An obvious (subdominant) source of *microwave radiation* is the *dust grains*, and their emission *might contribute to or modify the observed CMB anomalies*. The *zodiacal cloud* has been studied in detail for the *Planck* 2013 release and the *Planck* team in its 2015 analysis subtracted a fit to the *Kelsall model* for the *zodiacal cloud* before map making. The *Kelsall model* attempts to capture the *solar system dust emission* in the *infrared* and *microwaves* and is based on the analysis of COBE DIRBE observations.

When comparing the *Kelsall model* with *two meteoroid engineering models* (used by space agencies to reduce the hazard to launch a spacecraft into a shower of meteoroids), it has been found that those engineering models, depending on the chemical composition of the dust grains, *predict a much brighter zodiacal cloud at microwave frequencies*[102].

[102] Dikarev, V. V., & Schwarz, D. J. (December, 2015). Microwave thermal emission from the zodiacal dust cloud predicted with contemporary meteoroid models. *Astronomy and Astrophysics*, 584, A9; https://doi.org/10.1051/0004-6361/201525690.

The *Divine model* of the interplanetary meteoroid environment predicts meteoroid fluxes on spacecraft anywhere in the solar system from 0.05 to 40 AU from the Sun. This model uses data from micro-crater counts in lunar rocks from Apollo, meteor radar, and in situ measurements from Helios, Pioneer 10 and 11, Galileo and Ulysses. However, it does not make use of the infrared observations of COBE DIRBE. The *Interplanetary Meteoroid Engineering Model* (IMEM) and the *Divine model* use the *distributions in orbital elements and mass* rather than the *spatial density functions* of the *Kelsall model*, ensuring that the *dust densities* and *fluxes* are predicted in accord with *Keplerian dynamics* of the constituent particles in heliocentric orbits. IMEM is constrained by the micro-crater size statistics collected from the lunar rocks, COBE DIRBE observations of the *infrared emission* from the *interplanetary dust* at 4.9, 12, 25, 60, and 100 μm *wavelengths*, and Galileo and Ulysses *in situ* flux measurements.

The different predictions for *microwave emission* from the *solar system* of the three models is illustrated in figure 9.

To a first approximation, the *zodiacal dust foreground* produces a smooth band *along the ecliptic*, see figure 9. *This does not give rise to a hemispherical asymmetry, but it could cause alignments of low CMB multipoles with the ecliptic plane.* Additionally, it could contribute to a positive correlation at very large angular separations, as the *antipode* of a point close to the *ecliptic* is also close to the *ecliptic*. However, the shape of the emission from the *zodiacal cloud* cannot give rise to the type of alignment observed in the low ℓ *multipoles* of the *cosmic microwave background* (CMB) because it looks like a $Y_{\ell 0}$ in ecliptic coordinates, not $Y_{\ell \ell}$ (…). Therefore, while the *zodiacal dust* is unlikely to cause a lack of large angle correlation, it could change the significance of some of the anomalies.

Another solar system source of CMB *foreground* might be the *Kuiper belt* and more solar system related ideas have been studied in Frisch (2005)[110], proposing nearby interplanetary dust towards the nose of the *heliosphere* to be responsible of some of the unexpected alignments.

[110] Frisch, P. C. (October, 2005). Tentative identification of interstellar dust in the magnetic wall of the heliosphere. *Astrophys. J.*, 632, L143–6; https://iopscience.iop.org/article/10.1086/497909/pdf.

3.2. Milky Way

The next well-established layer of *foregrounds* are due to the Galaxy. At the highest frequencies *galactic thermal dust* is dominant and molecular lines from CO transitions contribute in various frequency bands.

Until recently it was believed that at *low frequencies synchrotron* and *free–free* emission are the dominant mechanisms. Interestingly enough, the Planck 2015 release overturned that point of view and showed that *free–free* and *spinning dust* are the dominant components at the *lowest frequencies*.

Shortly after the discovery of the low *multipole alignments*, one suspicion was that they could be caused by residual contamination due to *Galactic foregrounds*. If the Galactic plane signal contributes to the large-angle *cosmic microwave background* (CMB), multipole vectors should point in the direction of the plane and, more generally, alignments should be *Galactic* (and not largely *ecliptic*). This has been explicitly demonstrated by[114], who found that adding a Galactic-emission-shaped template contributing to the CMB map with an arbitrary weight does not lead to observed alignments.

[114] Copi, C. J., Huterer, D., Schwarz, D. J., & Starkman, G. D. (2006). On the large-angle anomalies of the microwave sky. *MNRAS*, 367, 1, 79–102; https://doi.org/10.1111/j.1365-2966.2005.09980.x.

The multipole vector analysis is particularly effective in this case, as it easily detects the directions singled out by the Galaxy, i.e. the *Galactic center* and the *Galactic poles*.

It also seems to be hard to explain a lack of correlation at the largest angular scales from residual galactic contamination. The Galaxy is quite close to us and thus highly correlated over large scales (it extends over much more than 60° on the sky). The comparison of the two-point correlation function for different masks, as well as constraining the analysis to correlations for which at least one point is close to the Galactic plane, are in full agreement with the hypothesis that foreground cleaned full-sky maps *do contain Galactic residuals that strongly affect the amount of correlation at the largest angular scales*. However, the fact that the most conservative masks provide the smallest amounts of correlation seems to indicate that *it is precisely the cleanest, most trustworthy regions on the sky that show the strongest evidence for the vanishing of angular correlations in the CMB*.

Recently it became clear that there is another type of *galactic foreground* that seems to be more *local* and might have a quite complex structure. The so-called *radio loops* are believed to be relics of *supernovae*. They have been detected at *radio frequencies* long ago, and have been believed to be of no relevance for the CMB *temperature anisotropies*. However, it was argued, especially in the context of *polarized* emission, that this might not be true. Most recently, it was shown that a *loop structure* in the vicinity of the *cold spot*, together with another structure called *radio loop I*, is able to almost perfectly reproduce the observed *quadrupole–octopole* map. If indeed these two structures would dominate the sky at the very *low multipole moments*, then the primordial fluctuations at those scales must be completely absent. This might explain the *alignments* and maybe to some extend the *dipolar modulation*, but *the lack of correlation would become more significant and would be in stark contrast to the* [*relativistic*] *ΛCDM model.*

3.3. Other foregrounds

There are a number of other *foregrounds* that could in principle have effect on, or even be the cause of, the anomalies. For example, the *local extragalactic environment* in form of *hot plasma* and a local Sunyaev–Zeldovich effect may play a role. However, these other *foregrounds* have not been studied in great detail in the context of the anomalies, partly because they are not thought to be able to generate features at *very large angular scales*.

The short summary of this section, therefore, is that while none of the aforementioned effects have been proven to cause any of the *cosmic microwave background* (CMB) anomalies, it is clear that these are physically well motivated *foregrounds* and an improved understanding of them will also help us to better understand the nature of the unexpected CMB features.

The major argument against a foreground related explanation of the CMB anomalies is the frequency independence of the observed anomalies. In fact, the anomalies show up at more or less the same statistical significance in four different *Planck* pipelines that lead to foreground cleaned maps of the full sky and in the corresponding WMAP pipeline. This implies that any so far unidentified foreground that would be responsible for one or all of the unexpected features of the CMB at large angular scales would have to mimic a *CMB fluctuation spectrum*, in order not to show up in the difference maps between the four foreground cleaned *Planck* maps (Commander, NILC, SEVEM, SMICA).

4. Cosmology

Perhaps the most exciting possibility is that some or all of the anomalies have a (common) *cosmological origin*. In this section, we consider a variety of proposed cosmological mechanisms whose manifestation could be the observed anomalies.

4.1. Kinetic effects

Earth's motion through the rest frame of the *cosmic microwave background* (CMB) leads to higher-order effects on the observed anisotropy, which could in principle affect conclusions about the observed *anomalies*. As already discussed above, these so-called *kinetic effects* have been studied for *low multipole moments* as well as for the *highest multipole moments* and both contribute to the final significance for the anomalies. *The kinetic effect on the quadrupole also provides another argument against a solar system, Galactic or local extra galactic foreground.* When this well-understood correction to the data is applied, evidence for the *alignments* becomes even stronger. *If those alignments were caused by, say, a Galactic foreground, the correct kinetic correction should be derived from the velocity of the solar system within the Galaxy and not with respect to the CMB frame.* In that case a 'wrong' kinetic correction would have been applied, which would be very unlikely to increase the *alignments* (a random correction actually leads most likely to a less significant alignment). This indicates that *the alignment is a physical effect and that it is not due to foregrounds.*

4.2. Local large-scale structure

Local structure—over/underdensities in the dark matter distribution within tens or few hundreds of megaparsecs of our location in the Universe—could in principle be responsible for some of the *alignments*. This class of explanation has a nice feature of producing large scale

246

effects relatively easily, since the small distance to us implies a large angle on the sky (see figure 2).

One possibility is the late-time *integrated Sachs–Wolfe effect* (ISW), or, in the non-linear regime the Rees–Sciama effect. This is the additional *anisotropy* caused by the decay of *gravitational potential* when the Universe becomes *dark energy-dominated* (*redshift* $z \leq 1$), and has a nice feature that it is achromatic. First estimates showed that the effect could give rise to the correct order of magnitude for the *quadrupole* and *octopole*. In … a single spherical structure was considered and it was argued that a single over or underdensity cannot give rise to the observed pattern. …

An argument against explaining the observed *alignments* with the *integrated Sachs–Wolfe effect* (ISW) is simply that it is *unlikely*: barring a suppression of *primordial* temperature fluctuations, the observed missing power at large angles generically requires a chance cancellation between the *local* ISW signal and the *primordial cosmic microwave background* (CMB) pattern. This is unlikely and, taken at face value, would imply another anomaly. Nevertheless, this idea can eventually be tested by means of cross correlation of the CMB maps with all-sky maps of the cosmic structure.

The idea that an unusually *large void* is in our vicinity has been revived repeatedly in the context of the *cold spot* anomaly. There had been a claim of an *underdensity* in the NVSS radio survey in the location of the CMB *cold spot*, but this feature was proven to not be statistically significant once the systematic errors in the survey, and in particular the known *underdensity* stripe in NVSS, are taken into account.

[The *NRAO VLA Sky Survey (NVSS)* was an astronomical survey of the Northern Hemisphere carried out by the *Very Large Array* (VLA) of the *National Radio Astronomy Observatory* (NRAO), resulting in an astronomical catalogue. The survey covers 82% sky, consisting of everything north of declination − 40 degrees. The observations were made in 'D' and 'DnC' configuration at

1.4 gigahertz (21 cm), with an angular resolution of 45 arcsec. Observations were made between September 1993 and October 1996.]

More recently, there was a claimed discovery of a large (~200 Mpc) *underdensity* ($\delta\rho/\rho \sim -0.15$) centered at *redshift* z ~ 0.2 in the distribution of galaxies in the 2MASS-wide infrared survey explorer (WISE) survey]. The *underdensity* lies in the direction of the CMB *cold spot*, leading to a fascinating possibility that the former is causing the latter via the *integrated Sachs–Wolfe effect* (ISW) effect. However, this causal explanation has been brought into question, as it appears that the *underdensity* is not sufficiently pronounced to cause the observed temperature *cold spot*. Future tests, discussed in section 5, will have a lot more to say about the *local structures* and their relation to *cosmic microwave background* (CMB) anomalies.

4.3. Primordial power spectrum—broken scale invariance

An *inflationary* scenario with the minimally short period of slow-roll, say just 50–60 e-folds, could accommodate breaking of the *scale invariance* during *inflation* at observationally accessible scales, that in turn could manifest itself as one or more of the CMB anomalies. Alternatively, one could also consider scenarios in which inflation deviates from its generic slow-roll behavior just 50–60 e-foldings before it ends. Many of those models find an improved quality of fit to the observations, however the improvement is typically not statistically significant given the additional parameters of the model. These models also make definite predictions on *tensor modes* and on the *polarization* of the *cosmic microwave background* (CMB) at the *largest angular scales*; thus, there are more handles that we can hope to exploit in the future. However, a generic problem of an explanation along these lines is that a new fine tuning (*why do we live in the epoch when the Universe is large enough to observe the first pre-inflationary scales*) is introduced in some of these models.

4.4. Primordial power spectrum—broken isotropy

Tests of *isotropy* and *homogeneity* of the initial conditions in the Universe have seen tremendous activity and development over the past decade. Two facts contributed to this. First, several of the anomalies, particularly the *parity anomaly* and the *hemispherical anomaly*, can be naturally modeled (and potentially explained) using modulation of either the *primordial temperature* or *primordial power spectrum*. Second, the advent of full-sky WMAP and *Planck* CMB maps enables the precise measurements required to constrain these models, particularly the high-ℓ *couplings* between the *multipoles*.

At the level of the CMB *temperature*, the *modulation* can most generally be written as

$$T(\hat{e}) = T_0 (\hat{e})[1 + f (\hat{e})] \tag{1}$$

where f is some function.

> [*Temperature modulation* in cosmology refers to the fine structure and anisotropy of the *cosmic microwave background* (CMB) radiation *polarization field*. It can be explained by *dipole modulation models*, which lead to correlations between different *multipoles*, similar to *temperature fluctuations*.]

Reference [59] studied both the *dipolar* and the *quadrupolar temperature modulation* (i.e. when f is proportional to Y_{1m} and Y_{2m}, respectively) in order to explain the *missing correlations at large scales* and the *quadrupole–octopole alignment*.

[59] Gordon, C., Hu, W., Huterer, D., & Crawford, T. (November, 2005). Spontaneous isotropy breaking: A mechanism for CMB multipole alignments. *Phys. Rev.*, D, 72, 103002; DOI: https://doi.org/10.1103/PhysRevD.72.103002.

While it is certainly possible to do so, Gordon, C., et al. (2005)[59] found that it is difficult to *naturally* arrange for the *missing correlations*, as it typically requires chance cancellations at large scales.

The discovery of *hemispherical asymmetry* gave much further impetus to the study of *modulations*, particularly the *dipolar* one written down in equation (9). It has become particularly interesting to ask what general mechanisms could produce such a *long-wavelength [temperature] modulation*. An *inflationary theory* could, in principle, accommodate models that produce *hemispherical asymmetry*, but such a model *would have to be multi-field and involve, for example, a large-amplitude superhorizon perturbation to the curvaton field*. It turns out that a *long wavelength, superhorizon-scale gradient in density* (the so called *Grischuk Zeldovich effect* could not produce the *dipolar asymmetry* starting with *adiabatic fluctuations* because the *intrinsic dipole* in the *cosmic microwave background* (CMB) produced by the *perturbation* is exactly canceled by the *Doppler dipole* induced by our peculiar motion. However, a *superhorizon isocurvature perturbation* could do the job. In that case, the effects of the *superhorizon fluctuation* would also presumably be seen as the *anisotropic distribution of large-scale structure on the sky, but no such effect has yet been detected in the distribution of quasars or galaxies and other tracers*.

An equally interesting possibility is that *anisotropic inflation* for a led to a breaking of statistical isotropy. The best-studied model posits that the *power spectrum* take the following form

$$P(k) = P(k) \, [1 + g_* \, (k \cdot d\hat{})^2], \tag{1}$$

where $d\hat{}$ is again a special direction. Such a model would imply that *inflation* was *anisotropic*, which could be caused by *coupling to vector fields, presence of magnetic fields* or *models motivated in supergravity*. First estimates of the parameter g_* found it is non zero at the huge significance of $\sim 9 \, \sigma$, but this was soon found to be due to *a known effect of asymmetric beams which had not been taken into account*. A later analysis based on *Planck* data gave the best constraint to date,

$g_* = 0.002 \pm 0.016$ at 68% C.L; the *Planck* team gets very similar constraints.

A *dipolar [temperature] modulation* may also result from a *parity-violating* exited initial state in the context of *slow-roll inflation*, a scenario that also predicts small *non-Gaussian* features.

It is also interesting that the apparent *breaking of statistical isotropy* can actually be an artifact of *non-Gaussianity*. More precisely, *coupling* between a *superhorizon long mode* and *shorter, observable modes* due to *primordial non-Gaussianity on superhorizon scales* can manifest itself in observations as a *preferred direction* on the sky. Essentially, one can thus 'trade' the breaking of statistical *isotropy* for the presence of primordial *non-Gaussianity*. ...

4.5. Topology

A *non-trivial topology* of the Universe might in principle both lead to a *lack of correlation at large angular scale* and introduce *alignments* and/or *asymmetries*, while preserving a *locally isotropic and homogeneous geometry*.

[The *shape of the universe* refers to both its *local* and *global geometry*. *Local geometry* is defined primarily by its *curvature*, while the *global geometry* is characterized by its *topology* (which itself is constrained by curvature). *General relativity* attempts to explain how *spatial curvature (local geometry)* is constrained by gravity. The *global topology* of the universe cannot be deduced from measurements of *curvature inferred from observations within the family of homogeneous general relativistic models alone*, due to the existence of *locally* indistinguishable spaces with varying *global topological characteristics*. For example; a multiply connected space like a 3 torus has everywhere zero curvature but is finite in extent, whereas a *flat simply connected space* is infinite in extent (such as Euclidean space). Current observational evidence (WMAP,

BOOMERanG, and *Planck* for example) imply that *the observable universe is spatially flat* to within a 0.4% margin of error of the *curvature density parameter* with an unknown *global topology*. It is currently unknown whether the universe is *simply connected* like Euclidean space or *multiply connected* like a torus. To date, no compelling evidence has been found suggesting the topology of the universe is not *simply connected*, though it has not been ruled out by astronomical observations.]

The idea is that the Universe might have a *finite size* which is not much larger than today's *Hubble distance*. In such a Universe there is a natural cut-off for *structures* at large scales and if the different large but compact dimensions are not of equal size, we could even imagine that a plane like the *quadrupole octopole plane* would be singled out.

Detailed studies of non-trivial topologies however did not find any statistically significant signal to substantiate those ideas. These included generic studies based on the circles-in-the-sky signature, which rule out with reasonable confidence that there is a non-trivial closed loop in the Universe with length less than 98.5% of the diameter of the *last scattering surface*. Searches for anomalous correlations in the *cosmic microwave background* (CMB) can extend these bounds for specific *manifolds* or classes of *manifolds* to slightly larger distances.

For non-trivial *topology* to be observable, whether or not it is behind any of the *mysterious large-scale features* in the *cosmic microwave background* (CMB), the characteristic length scale of the fundamental domain would need to be comparable to the *Hubble distance*. We would then be faced with another coincidence problem—*why do we live in an epoch in which we are able to see a non-trivial topology?*

A short summary of the *cosmological explanations* offered so far is that the ideas mentioned here could all explain at least one 'anomaly atom', however, *none of them has been demonstrated to be detectable at a statistically significant level*. The *kinetic effects* must be taken into account, but *do not seem to explain any of the new features of the CMB*.

The *local large-scale structure* exists and must be better understood — especially from non-CMB observations (see below). *Primordial physics might lead to broken scale invariance, very large non-adiabatic modes and non-trivial topologies.* All three suffer from a coincidence problem. This problem might be most prominent for finite topologies, like a 3-torus. When combined with inflation such a solution requires a single short period of *inflation.* All ideas of *primordial power suppression* suffer from a regeneration of large-scale power via the *integrated Sachs–Wolfe effect* (ISW) effect at late times. This aspect is tamed in finite topologies, but in that case the alignment of modes is diluted via the ISW effect. Finally, the idea *to break the isotropy of the primordial power spectrum* seems—when compared to data—not to be implemented in the *cosmic microwave background* (CMB) *on all scales* (in particular *there is no evidence for such an effect at $\ell \geq 60$*) and thus must be combined with a breaking of *scale invariance.* On top it is unclear how such a *primordial breaking of isotropy* would generate a *lack of correlation at large angular scales.*

5. Way forward and conclusion

The origin and nature of *cosmic microwave background* (CMB) *temperature anomalies* can be tested with other observations in cosmology. The best way to proceed along those lines is to assume the model for the CMB *temperature anomalies*, and test that model with other data.

The simplest model that can be tested is that the anomalies are simply *fluke events in the standard [relativistic ???] ΛCDM cosmological model.* In the language of frequentist statistics, the anomalies in this case are realized in a very small fraction of random realizations within the underlying ΛCDM model. Then one could simply ask how the other observations, beyond CMB *temperature*, are affected as seen in this small subsection of anomalous ΛCDM realizations. One could similarly test other, non-ΛCDM models for the anomalies whether they are fundamental or purely phenomenological and, if desired, use Bayesian statistics as well.

We now discuss how other observations in cosmology can help understand the CMB *temperature anomalies*.

5.1. CMB polarization

Any given model for the *cosmic microwave background* (CMB) *temperature anomalies* has predictions for *polarization*. Given the increasingly precise polarization measurements, it is of great interest to make predictions for polarization under different models for the anomalies. *The polarization data from WMAP and Planck are good enough to produce reliable polarization power spectra, but are not high enough signal-to-noise to produce maps of the polarization field—* especially not at *very large scales* where *foregrounds* are extremely difficult to remove. Nevertheless, new generations of experiments, such as LiteBird, CORE, and CMB-S4 hold promise for accurate maps of *polarization* down to the lowest *multipoles*.

First predictions for *polarization* were carried out by Dvorkin, Peiris, & Hu, (2008)[177] who studied how a *dipolar modulation model* that can explain the *hemispherical asymmetry* and a *quadrupolar modulation model* that can explain the *quadrupole–octopole alignment* can be constrained with *polarization* map data.

[177] Dvorkin, C., Peiris, H. V., & Hu, W. (March, 2008). Testable polarization predictions for models of CMB isotropy anomalies. *Phys. Rev.*, D, 77, 063008; https://doi.org/10.1103/PhysRevD. 77.063008.

For the *dipolar* case, they showed that predictions for the correlation between the first 10 *multipoles* of the *temperature* and *polarization* fields can typically be tested at better than the 98% CL; while for the *quadrupolar* case, predicted correlations between *temperature* and *polarization multipoles* out to $\ell = 5$ provide tests at the 99% CL or stronger. This was followed up by [178] who assumed that the *suppressed correlations* observed in WMAP and *Planck temperature* data are just *unlikely realizations in the [relativistic] ΛCDM model*, and studied what that hypothesis predicted for CMB *temperature-polarization* cross correlation.

[178] Copi, C. J., Huterer, D., Schwarz, D. J., & Starkman, G. D. (August, 2013). Large-angle cosmic microwave background suppression and polarization predictions. *MNRAS*, 434, 4, 3590–6; https://doi.org/ 10.1093/mnras/stt1287.

Their conclusion is that while the temperature-polarization cross-correlation cannot definitively be expected to be able to rule out ΛCDM, one can nevertheless construct statistics that have a good chance (~50%) of excluding the hypothesis at a high statistical confidence (>3 σ). Similar results were obtained for the predicted utility of large-angle temperature-lensing-potential correlation function measurements; however, in [180] the authors showed that the *large-angle polarization auto-correlation* function may be more promising.

[180] Yoho, A., Aiola, S., Copi, C. J., Kosowsky, A., & Starkman, G. D. (March, 2015). Microwave background polarization as a probe of large-angle correlations. *Phys. Rev.*, D, 91, 123504; https://doi.org/ 10.1103/PhysRevD.91.123504.

The latter also began exploring the alternative hypothesis that the lack of *large angle temperature auto-correlation* reflects a lack of large-distance correlations in the metric potentials, and note that the *large angle polarization auto-correlation* appears well suited to test this hypothesis.

5.2. Large-scale structure

The *distribution of galaxies*, or *other tracers of the large-scale structure*, may be particularly useful to clarify the nature of *at least some of the observed cosmic microwave background* (CMB) *anomalies. Galaxies* have now been mapped over the whole sky out to z ~ 0.2, e.g.

by the 2MASS and WISE surveys, and to a depth of z ~ 0.7, by the SDSS, including a subsample of more than two million *galaxies* and *quasars* that have spectroscopic information. In the future, a combination of the Dark Energy Survey, Dark Energy Spectroscopic Instrument, Euclid, Large Synoptic Survey Telescope (LSST) and Wide-Field Infrared Survey Telescope will map out the *galaxy distribution* over the whole sky out to *redshift* beyond one, while eROSITA will carry out a complete census of *x-ray clusters* in the observable Universe. Other tracers, such as *quasars* and *radio galaxies*, are particularly useful since they are at *redshifts* of a few and probe an even larger volume. Eventually H_I surveys, e.g. by means of the square kilometer array (SKA), will map all *galaxies* containing *neutral hydrogen* out to a redshift of a few. The SKA will also allow us to eventually pin down the *cosmic radio dipole* at per cent level and thus provide a precise test of the *proper motion hypothesis* of the CMB *dipole* and possibly identify a significant *structure dipole*.

To give another example, *missing power on large angular scales leads, assuming [the relativistic] ΛCDM, to corresponding suppression of power on a gigaparsec scale* (that is, the *power spectrum of matter fluctuations* P(k) is suppressed for $k \leq 0.001$ h Mpc^{-1}). It turns out that a future very large volume survey, such as LSST, could in principle measure power on such large scales to a sufficient accuracy to detect the purported *suppression in power* at a statistically significant level.

The *cold spot* offers a particularly appealing target for the large-scale structure surveys because of its smaller *angular size*. While the aforementioned evidence for the *underdensity in the large-scale structure* is inconclusive (see section 4.2), the aforementioned future wide, deep surveys will offer a fantastic opportunity to correlate the large-scale structure to the *cosmic microwave background* (CMB) anomalies.

5.3. Foregrounds

From the discussion in section 3, *it is clear that both solar system and galactic foregrounds need to be better understood*. For example, we already mentioned that *three different models for the zodiacal dust emission mutually disagree by more than an order of magnitude*. It would be important to investigate the prospect of dedicated observations targeted specifically in regions where the *zodiacal dust signal* becomes more important to test those models. Similarly, *the nature of spinning dust is not well understood* (see e.g. [194,195]).

[194] Draine, B. T., & Hensley, B. S. (2012). The Submm and mm Excess of the SMC: Magnetic Dipole Emission from Magnetic Nanoparticles? *Astrophys. J.*, 757, 103; https://doi.org/10.1088/0004-637X/757/1/103

[195] Hensley, B. S., Draine, B. T., & Meisner, A. M. (2015). A Case Against Spinning PAHs as the Source of the Anomalous Microwave Emission. arXiv:1505.02157; https://arxiv.org/abs/1505.02157.

The same holds for *the extrapolation of known radio loops into the relevant microwave bands*. In both cases, *dedicated radio and microwave observations would be useful*. The DeepSpace project in Greenland is planning to cover those frequency bands. More generally, other astrophysical observations in the coming 10–20 years should be able to provide significant new information about foregrounds and their effects on the *cosmic microwave background* (CMB) anomalies.

6. Executive summary

We summarized the evidence that several *cosmic microwave background* (CMB) anomalies are real features. We identify three anomaly 'atoms': the *lack of large angle correlation*, the *mutual alignment of the lowest multipole moments*, and the *hemispherical asymmetry*. These three 'atoms' seem to be orthogonal and independent of each other in the realm of the minimal *inflationary* [*relativistic*] ΛCDM model. Any proposal to explain (or just parameterize) these new *cosmic microwave background* (CMB) features should address at least two of the 'atoms' as one of them might, after all, still be a statistical fluctuation. *In an effort to find a compelling explanation, several new theoretical ideas have been considered, covering a broad spectrum from the physics of dust grains to exotic theories of the very early Universe.* Currently, *the physics behind the CMB anomalies is still unknown*, but new observations of the CMB (especially of *polarization*) and new observations at other wavebands, both of the *large-scale structure* and of potential *foregrounds*, will provide significant new information and provide us with powerful new tools to eventually resolve the puzzle of the anomalies.

Ćirković, M. M., & Perović, S. (2018). Alternative explanations of the cosmic microwave background: A historical and an epistemological perspective.

Studies in History and Philosophy of Science, Part B: Studies in History and Philosophy of Modern Physics, 62, 1–18; arXiv:1705.07721; https://doi.org/10.1016/j.shpsb. 2017.04.005.

Highlights

• The orthodox explanation of *cosmic microwave background* (CMB) has developed in four stages.
• A variety of *relativistic* and *non-relativistic* alternatives challenged it.
• Theoretical debates and evidence incrementally constrained the alternatives.
• The epistemological status of evidence in cosmology is complex.
• Various aspects of alternative explanations of CMB may be fruitful.

———————————

Abstract

We historically trace various non-conventional explanations for the origin of the *cosmic microwave background* (CMB) and discuss their merit, while analyzing the dynamics of their rejection, as well as the relevant physical and methodological reasons for it. It turns out that there have been many such unorthodox interpretations; not only those developed in the context of theories rejecting the *relativistic* ("*Big Bang*") paradigm entirely (e.g., by Alfvén, Hoyle and Narlikar) but also those coming from the camp of original thinkers *firmly entrenched in the relativistic milieu* (e.g., by Rees, Ellis, Rowan-Robinson, Layzer and Hively). In fact, the orthodox interpretation has only incrementally won out against the alternatives over the course of the three decades of its multi-stage development. While on the whole, none of the alternatives to the hot *Big Bang* scenario is persuasive today, we discuss the epistemic ramifications of establishing orthodoxy and eliminating alternatives in science, an issue recently discussed by philosophers and historians of science for other areas of physics. Finally, we single out some plausible and possibly fruitful ideas offered by the alternatives.

Introduction

The discovery of the *cosmic microwave background* (CMB) in 1965 by Arno Penzias and Robert Wilson and interpreted by Robert H. Dicke and his co-workers was a turning point in 20th century cosmology. It divided cosmology into an epoch of sometimes heated cosmological controversy (Kragh, 1996) and an epoch of solidified support for the standard cosmological paradigm, popularly known as the hot *Big Bang* cosmology (Peebles, Page, & Partridge, 2009). Actually, attributing the discovery of the CMB to Penzias and Wilson is a bit misleading, first, because they were not looking for it and, second, because it had been predicted by Gamow and his collaborators a few decades earlier.

They initially interpreted the accidentally detected signal as a noise caused by an artefact; they were not aware it had anything to do with a physical phenomenon of the utmost importance for cosmology. Their detection of the signal had far-reaching implications, however, not least of which was a now overlooked interpretation race in which they themselves did not participate.

The fact that the 1965 discovery was a clear watershed creates the impression of inevitability of the currently standard interpretation of the great CMB discovery as a remnant of primordial fireball, and that no alternative interpretations have been offered, seriously or half-seriously, by distinguished cosmologists. The impression of the inevitability of the current view is shared by astronomers and laypersons alike. Two of the best cosmology textbooks available, by Coles and Lucchin (1995) and Peacock (1999), reinforce this impression. Peacock even notes, with a poetic flourish, "The fact that the properties of the *last-scattering surface* are almost independent of all the unknowns in cosmology is immensely satisfying, and gives us at least one relatively solid piece of ground to act as a base in exploring the trackless swamp of cosmology" (p. 290).

From the point of view of the astrophysics community, the validity of the orthodox interpretation of *cosmic microwave background* (CMB) is largely resolved, with some doubts voiced from time to time (e.g., Baryshev, Raikov & Tron, 1996). And as far as the general issue of the choice of cosmological models is concerned, the standard cosmological model seems to rest on a secure foundation (for review of some exotic alternatives, see Ellis, 1984).

Yet López-Corredoira (2014) has quite recently examined some alternative cosmological models from a sociological point of view. This is important, as the emergence of alternatives and their destiny is a complex issue at the heart of scientific knowledge production and the discovery process. For instance, Cushing (1994) argues that a perfectly viable alternative to the Copenhagen interpretation of Quantum Mechanics, Bohm's mechanics, has been side-lined because it was devised later on. And Chang (2009), Chang (2010) says forgotten and abandoned alternatives are often alternate routes to discoveries that were never taken. He demonstrates this using relevant examples in chemistry. Perovic (2011) analyses how subtle changes in experimental conditions influence the possibility of emerging and often crucial alternative theoretical accounts in particle physics, while Dawid, Hartmann, and Sprenger (2015) offer a Bayesian analysis of theoretical preferences when viable theoretical alternatives are not available.

The CMB is another case, and in many respects, a different and fruitful case, the study of which can enrich this strand of methodological and philosophical research. Generally speaking, in the scientific fields that reconstruct evidence from *observations*, the epistemic standing of orthodox thought is tied to the epistemic standing of available alternatives. Evidence in such cases is, on the whole, very different from evidence provided in, say, experiments in solid state physics, in the sense that the underdetermination of theoretical accounts by evidence is bound to be much more pronounced and longer lasting. The wiggle room for alternative interpretations is much wider in a field such as cosmology than in experimental physics, as the latter provides much

261

more direct evidence in debates and thus severely constrains theoretical accounts of relevant phenomena. The *cosmic microwave background* (CMB) was a milestone discovery, but it would be misleading to think it played a role identical to that, for instance, played by the evidence delivered by a particle collider in competing theoretical approaches to the existence of an elementary particle. Its role unravelled much more gradually.

Given this, it is wise to avoid treating side-lined alternative interpretations in the same way as we justifiably would experimentally falsified alternatives in experimental physics. Instead, we should generally regard them as a resource that can potentially be revised and revived (despite occasional fairly straightforward falsifications of its certain aspects) The evidence of orthodoxy does not necessarily justify our outright discarding of the alternatives in cosmology. In fact, establishing orthodoxy may unjustifiably boost the CMB's epistemic standing by eliciting ignorance or a too-hasty dismissal of the existing alternatives, in part by propagating an inadequate history of the field and systematically, albeit unjustifiably, downplaying existing alternatives. Failing to understand the subtleties of the history of how orthodox thought about CMB was established runs the risk of generating widespread prejudice that opinions dissenting from the standard paradigm are both few and insignificant.

In short, the CMB provides an incentive for philosophically minded historical research. Just how convincing was the account that became the standard CMB interpretation in the first years after Penzias' and Wilson's discovery or during the first decade or two thereafter? Were any viable alternatives neglected at the time? How convincing is the account currently, and are there any viable alternatives now? Has there been enough critical examination in the modern practical work on the issue? All these questions are part of the complex and insufficiently studied problematic of paradigm formation in modern cosmology (Kragh, 1997; Norton 2017). In the first part of the paper (2 The *microwave background* phenomenology and its standard interpretation,

3 *Moderate unorthodoxies*: CMB as a relic of Population III objects, 4 *The CMB in radical unorthodox models* (without *Big Bang*)), we offer a historical case study of the formation of the alternatives in modern cosmology, setting the basis for an assessment of their respective epistemic standing in the second part of paper (5 *The formation of the orthodoxy and the alternatives*: an epistemological framework, 6 *What of alternatives?*).

Peebles (1999) commentary on the centennial re-edition of Penzias and Wilson (1965) paper is a good starting point for our research into the historical context of the CMB: A willingness to believe such an elegant gift from nature surely also played a significant role in the early acceptance of the CBR [*cosmic background radiation*] interpretation… During four decades of involvement with this subject, I have grown used to hearing that such advances have at last made cosmology an active physical science. I tend to react badly because I think cosmology has been an active physical science since 1930, when people had assembled a set of measurements, a viable theoretical interpretation, and a collection of open issues that drove further research. This equally well describes cosmology today.

This comment sets the stage for the article. The "willingness to believe" the standard model and a lack of confidence in the seriousness of the pre-1965 cosmological research are key ingredients in the standard, streamlined view of the history of physical cosmology. (Peebles 2014) There is a widespread impression that the microwave noise detected serendipitously by Penzias and Wilson threw us into an epoch of serious, quantitative cosmology and that the essential validity of the hot *Big Bang* paradigm has remained unchallenged ever since. As Coles and Lucchin (1995) suggest, "it is reasonable to regard this discovery as marking the beginning of 'Physical Cosmology'" (p. xiii).

Yet the impression is wrong and creates a false picture of both the history and the methodology of cosmology. The facts about multiple methodologically sound alternative explanatory hypotheses of the *cosmic microwave background* (CMB) are mostly forgotten. Consequently, important historic-philosophical lessons about contemporary cosmological research are missed, and a source of potentially valuable ideas side-lined. It is worth trying to weave a historical tapestry of this admittedly amazing development by considering some strands presently deemed peripheral. The general motivation for this study is perhaps best expressed by Kragh's (1997) comments on the history of cosmology: There is a tendency to streamline history and ignore the many false trails and blind alleys that may seem so irrelevant to the road that led to modern knowledge. It goes without saying that such streamlining is bad history and that its main function is to celebrate modern science rather than obtain an understanding of how science has really developed. The road to modern cosmology abounded with what can now be seen were false trails and blind alleys, but at the time were considered to be significant contributions.

The story of the CMB alternative interpretations is paradigmatic in this respect. Many scientists and popularizers of science, perhaps justifiably, use every opportunity to hail the orthodox interpretation of CMB as one of the greatest, often as the greatest triumph of modern cosmological science. Yet its' often-professed role in terminating the cosmological controversy blurs the distinction between the physical phenomenon and the historical role of the dominant interpretation, ascribing some form of "progressive" value to the CMB photons themselves. The necessary palliative is certainly the study of the non-standard, minority interpretations which challenged the prevailing orthodoxy. In addition, as frequently happens in such circumstances, alternative theories may contain valuable side ideas, motivations, and conjectures. Because these theories are usually regarded as failures, their insights are understandably overlooked. This is actually quite common in the history of physics. For instance, in the cases of Machian *theories of gravitation*, such as Brans-Dicke theory (e.g., Dicke, 1962) or Wheeler-Feynman *action-at-a-distance classical electrodynamics* (Wheeler & Feynman (1945), Wheeler & Feynman (1949), Hoyle & Narlikar (1964), Hoyle

264

& Narlikar (1971), Hogarth (1962)), we encounter concepts too radical for their epoch, but which have since become the focus of debates in inflationary cosmology or of philosophical discussions on the arrow of time (Linde (1990), Price (1991)). Such cases offer an additional pragmatic argument for studying well-motivated unorthodoxies in their own right.

Finally, from a broader point of view of criticisms of cosmology in general, a constant feature of 20th century science (e.g., Dingle, 1954; Disney, 2000), the issue of the epistemological significance of the *cosmic microwave background* (CMB) is still important. If we now, post-CMB, consider ourselves entitled to high-precision models and predictions for the physical state of the early universe and corresponding traces and relics, what methodological desiderata do we use to derive such predictions? What supports our extrapolating to the states of matter many orders of magnitude more extreme than anything we encounter in a laboratory? To answer these and similar questions, we need to shed light on the emergence and acceptance of the standard CMB interpretation and the rejection of the alternatives.
…

Section snippets:

The microwave background phenomenology and its standard interpretation

The standard interpretation of CMB as a remnant of the primordial fireball was suggested by Dicke and his coworkers in their 1965 seminal paper (Dicke, Peebles, Roll & Wilkinson, 1965). While early predictions of Gamow and his students should not be discounted, the true history of the physics behind the CMB begins with Doroshkevich and Novikov (1964) and Dicke et al. (1965); in short, this was a set of ideas whose "time had come" and, hence, was taken seriously by both Soviet and American …

Moderate unorthodoxies: CMB as a relic of Population III objects

The prime examples of moderate unorthodoxies are developed within the cosmological models of cold or tepid *Big Bang*. Such models are variations on the standard theme of the singular origin of the universe in *relativistic Friedman models*, but under different initial conditions, particularly a low or intermediate value of the *photon-to-baryon* (or entropy per baryon) *ratio* η. As we discuss below, these models have some conceptual advantages over the standard hot (= high value of $\eta \sim 10^9$) *Big Bang*. …

The CMB in radical unorthodox models (without Big Bang)

The *classical steady-state theory* of Bondi, Gold and Hoyle (Bondi and Gold 1948) was very much alive at the time of the discovery of the *cosmic microwave background* (CMB). The theory already had several distinct problems, mostly with the radio source counts, as well as recently discovered high-*redshift* objects, QSOs, but these obstacles did not seem insurmountable. An excellent account in a monograph by Kragh (1996) shows how the *steady-state* paradigm had managed to overcome seemingly serious observational refutations, like …

The formation of the orthodoxy and the alternatives: an epistemological framework

We have seen several interesting trends in the dissenting tradition of CMB origin. In the incrementally emerging mainstream cosmology, i.e., the hot *Big Bang* model and a corresponding CMB interpretation (see Section 1), in the first few decades after the discovery, no individual "heresy" came close to attracting wide attention. The most seriously debated unorthodox solution to the CMB origin puzzle is likely Rees' model (1978). It is useful, as it contains most of the alternative ideas and …

What of alternatives?

The story of the *cosmic microwave background* (CMB) origin offers insights into the nature of the progress of modern science – its good and bad points alike. The role of the empirical but unexpected discovery of the CMB as unravelling the deepest mysteries of the origin of the universe was immediately and widely recognized by almost the entire cosmological community, including most researchers with unorthodox views. In general, it helped persuade a large portion of the wider scientific community that cosmology is a serious, … .

Handley, W. (February, 2021). Curvature tension: evidence for a closed universe.

Phys. Rev., D, 103, L041301; https://doi.org/10.1103/PhysRevD.103. L041301; also at arXiv e-prints; https://arxiv. org/pdf/1908.09139.

[1] Astrophysics Group, Cavendish Laboratory, J. J. Thomson Avenue, Cambridge, UK.

[2] Kavli Institute for Cosmology, Madingley Road, Cambridge, UK.

[3] Gonville & Caius College, Trinity Street, Cambridge, UK.

This work was performed using resources provided by the Cambridge Service for Data Driven Discovery (CSD3) operated by the University of Cambridge Research Computing Service, provided by Dell EMC and Intel using Tier-2 funding from the Engineering and Physical Sciences Research Council (capital grant EP/P020259/1), and DiRAC funding from the Science and Technology Facilities Council.

Abstract

The *curvature parameter tension* between *Planck* 2018, *cosmic microwave background lensing*, and *baryon acoustic oscillation data* is measured using the suspiciousness statistic to be 2.5 to 3 σ. Conclusions regarding the *spatial curvature of the universe* which stem from the combination of these data should therefore be viewed with suspicion. Without *cosmic microwave background* (CMB) *lensing* or *Baryon acoustic oscillations* (BAO), *Planck* 2018 has a moderate preference for *closed universes*, with Bayesian betting odds of over 50:1 against a *flat universe*, and over 2000:1 against an *open universe*.

[*Baryon acoustic oscillations* (BAO) are fluctuations in the density of the visible *baryonic matter* (normal matter) of the universe, caused by acoustic density waves in the primordial *plasma* of the *early universe*. In the same way that *supernovae* provide a "standard candle" for astronomical observations, BAO *matter clustering* provides a "standard ruler" for length scale in [*Big Bang*] cosmology.

According to *Einstein's theory of relativity*, under which space itself can curve, the universe can take one of three forms: *closed* like a sphere, or *open* with negative curvature like a saddle, or *flat* like a sheet of paper (but in 3 dimensions). According to this theory the universe's *density* determines whether it is *closed*, *open*, or *flat*. In a *closed universe* the universe's density is great enough for its gravity to overcome the force of expansion, then space-time is finite and expansion eventually stops and the universe begins to contract and curved back on itself. If you were to travel in a straight line in any direction, you would eventually return to your starting point. In an *open universe,* on the other hand, the universe's density is low and unable to stop the expansion, then space will warp in the opposite direction and the universe continues to expand indefinitely, with no end in sight. In a *flat universe* the density is such that it expands in every direction without curving positively or negatively. A *curved universe* corresponds to Einstein's theory of *General Relativity*; a *flat universe* corresponds to his theory of *Special Relativity*.]

INTRODUCTION

Quantifying the consistency between different cosmological datasets has become increasingly important in recent years. *Observations of the early and late time universe give undeniably inconsistent predictions for the present-day value of the expansion rate*[1], a phenomenon known as the *Hubble tension.*

[1] Verde, L., Treu, T., & Riess, A. G. (September, 2019). Tensions between the early and late Universe. *Nat. Astron.*, 3, 891–5; https://doi.org/10.1038/s41550-019-0902-0.

This paper explores a second curvature tension present [in models based on Einstein's theory of relativity] within early-time observations. Despite the long history of [*general relativistic*] cosmological models which include *spatial curvature*, in the modern era *there is a strong research community bias toward a flat universe.* This is partly on theoretical grounds, since *the prevailing (and very successful) inflationary theory of the primordial universe predicts a late-time cosmos which is extremely close to flat.* Curved models are also likely under-represented in the literature due in large part to the increased theoretical and numerical computational cost associated with curved cosmologies*.

> * The results presented in this paper required over twelve years of high-performance computing time (or two weeks on 320 cores).

However, most of the bias toward *flat cosmologies* is undoubtedly derived from *observational* considerations, which ever since COBE and BOOMERANG have confirmed that the universe is consistent with a *flat cosmology* to within $\Omega_K = \pm 10\%$. *The primary conclusion of this paper is that such a position can no longer be consistently held*, as *Planck* 2018 *cosmic microwave background* data[11,12] alone suggests a model [*based on Einstein's theory of General Relativity*] that is *closed* at $\Omega_K \sim -4.5\% \pm 1.5\%$, with betting odds of 50:1 against a *flat universe*.

> [11] Planck Collaboration. (July, 2018). *Planck 2018 results*. VI. Cosmological parameters. arXiv e-prints, art.arXiv:1807.06209.
> [12] Planck Collaboration. (July, 2019). *Planck 2018 results*. V. CMB power spectra and likelihoods. arXiv e-prints, art.arXiv:1907.12875.

Other datasets such as *Planck* 2018 *cosmic microwave background* (CMB) *lensing* or *baryon acoustic oscillations* which strongly suggest a *flat universe* are quantitatively *inconsistent* at 2.5 to 3 σ *with CMB data alone*, and should not be confidently combined until this tension is released. Results are visualized in Fig. 1 and summarized in Figs. 2 and 3.

BACKGROUND

Cosmology with curvature (KCDM)

Under the extended Copernican principle, the universe is assumed at zeroth order to be *homogeneous* and *isotropic* at the *largest scales. Einstein's theory of General Relativity under these assumptions yields the Friedmann Lemaitre Robertson Walker cosmology.* The metric in reduced spherical polar coordinates takes the form

$$ds^2 = dt^2 - a(t)^2 \left[dr^2/(1 - Kr^2) + r^2(d\theta^2 + \sin^2\theta \, d\varphi^2) \right] \quad (1)$$

whereby *homogeneous* and *isotropic* spatial slices expand or contract over cosmic time t via the *scale factor a.* The spatial slices come in one of three forms, *flat Euclidean space* (K = 0), *open hyperbolic space* (K = − 1) or *closed hyper spherical space* (K = + 1). Curvature is typically quantified via its fractional contribution to the *cosmic energy budget* today Ω_K, which is measured as a percentage and has the opposite sign to K. In this work, I follow the notation of Handley, (2019)[18] and denote the *concordance cosmology* with K = 0 as ΛCDM, and its extension with Ω_K considered as an unknown parameter as KΛCDM.

[18] Handley, W. (December, 2019). Primordial power spectra for curved inflating universes. *Phys. Rev.* D, 100, 123517; https://doi.org/10.1103/PhysRevD.100.123517

Bayesian statistics

Given a set of data D and a predictive model M with parameters θ, Bayes' theorem relates the statistical inputs to inference (the likelihood and prior) to the statistical outputs (the posterior and evidence):

$$P(D| \theta, M) \times P(\theta| M) = P(\theta| D, M) \times P(D| M)$$
Likelihood Prior = Posterior x Evidence $\quad (2)$
$$\mathscr{L} \times \pi = P \times Z$$

The *posterior* is the central quantity in parameter estimation (what the data tell us about parameters of a model), whilst the *evidence* is pivotal in model comparison (what the data tell us about the relative quality of a model). The *evidence ratio* gives the *Bayesian betting odds* one would assign between two competing models, assuming equal *a priori* model *probability*. Parameter estimation is typically performed by compressing the high-dimensional *posterior* into a set of representative samples via a *Markov-Chain Monte-Carlo process*. The *evidence* is derived from the *likelihood* by marginalizing over the *prior*

$$Z = \int \mathscr{L}(\theta)\pi(\theta) \, d\theta = <\mathscr{L}>_\pi, \qquad (3)$$

which may be computed numerically via a *Laplace approximation, Savage Dickey ratio, nearest neighbor volume estimation* or *nested sampling*. If one has access to the *evidence*, then it is straightforward to compute the *Kullback-Leibler divergence*

$$D = \int P(\theta) \log [P(\theta)/\pi(\theta)] \, d\theta = < \log P/\pi >_P, \qquad (4)$$

which quantifies the *degree of compression* from *prior* to *posterior* provided by the data.

As a model comparison tool, the *evidence* naturally quantifies Occam's razor, *incorporating a parameter volume-based penalty that penalizes models with unnecessary constrained parameters*. This penalty factor may be approximated using the difference in KL *divergence* between models. For further detail on Bayesian statistics, readers are recommended references[22,32-34].

[22] MacKay, D. J. C. (2003). *Information Theory, Inference and Learning Algorithms*. Cambridge University Press, Cambridge.
[32] Hobson, M. P., Jaffe, A. H., Liddle, A. R., Mukherjee, P., & Parkinson, D. (2010). *Bayesian Methods in Cosmology.*
Bayesian Methods in Cosmology. Cambridge University Press, Cambridge.
[33] Trotta, R. (January, 2017). *Bayesian Methods in Cosmology*. arXiv e-prints, art. arXiv:1701.01467.

[34] Sivia, D. & Skilling, J. (2006). *Data Analysis: A Bayesian Tutorial.* Oxford science publications. Oxford University Press, Oxford.

Tension quantification

The *Bayesian evidence* from Eq. (3) also appears in *tension quantification*. The *Bayes ratio*

$$R = P(D_2|D_1)/P(D_2) = P(D_1|D_2)/P(D_1) = Z_{12}/Z_1 Z_2 \quad (5)$$

quantifies the compatibility of two datasets D1 and D2 by giving a *Bayesian's relative confidence* in their combination. As with many Bayesian quantities, R is naturally *prior* dependent. In the absence of well-motivated *priors*, or if *deliberately over-wide ranges on parameters are used* (as is often the case in cosmology), the *prior* dependency can be removed from R using the KL *divergence* from Eq. (4). Dividing R by the *information ratio* I gives the *suspiciousness* S

$$S = R/I, \qquad \log I = D_1 + D_2 - D_{12} \quad (6)$$

Using a Gaussian analogy, one may calibrate S into a *tension probability* p and convert this into an equivalent "sigma value" via the *survival function* of the chi-squared distribution

$$p = \int_{d-2\log S}^{\infty} \chi_d^2(x)\, dx, \qquad \sigma = \sqrt{2}\ \mathrm{Erfc}^{-1}(p), \quad (7)$$

where Erfc^{-1} is the *inverse complementary error function* and d is quantified by the *Bayesian model dimensionality*

$$d^- = d^-_1 + d^-_2 - d^-_{12}, \qquad d^-/2 = \langle(\log \mathscr{L})^2\rangle_P - \langle\log \mathscr{L}\rangle^2_P \quad (8)$$

The differences in model dimensionalities recover the number of shared constrained parameters. See [35,36] for a more detailed discussion of tension and model dimensionality.

[35] Handley, W., & Lemos, P. (August, 2019). Quantifying tensions in cosmological parameters: Interpreting the Dark Energy Survey

(DES) evidence ratio. *Phys. Rev.*, D, 100, 043504; https://doi.org/10.1103/PhysRevD.100.043504.

[36] Handley, W., & Lemos, P. (July, 2019). Quantifying dimensionality: Bayesian cosmological model complexities. *Phys. Rev.*, D, 100, 2, 023512; https://doi.org/10.1103/PhysRevD.100.023512.

Other tension metrics are also available.

In this work the *triple tension* is also introduced to quantify the *tension* between three datasets simultaneously, extending the definition of R, I and d from Eqs. (5), (6) and (8) when necessary to

$$R = Z_{123}/Z_1 Z_2 Z_3, \qquad \log I = D_1 + D_2 + D_3 - D_{123}$$
$$d = d_1 + d_2 + d_3 - d_{123} \qquad (9)$$

with obvious generalization to four or more parameters.

METHODOLOGY

The observational datasets used for this analysis are described in Tab. I. Note that throughout this paper *lensing* is an abbreviation for *"Planck CMB lensing"*.

Bayesian *evidences* and *posteriors* are computed using *nested sampling* provided by CosmoChord, an extension of CosmoMC using PolyChord to sample efficiently in high dimensions and exploit the fast slow cosmological parameter hierarchy. The post processing computations for the evidence, KL divergence and Bayesian model dimensionality detailed in Figs. 2 and 3, as well as the posterior plots in Figs. 1 and 4 are computed using the anesthetic software package.

All cosmological and nuisance parameters are varied in the nested sampling runs. The *cosmological prior widths* (visualized in Fig. 4) are narrower than the CosmoMC defaults. This is necessary since nested sampling begins by exploring the deep tails of the distribution. In these regions unphysical and unforeseen combinations of parameters causes cosmology codes to fail. Whilst *flat cosmology* codes have been extensively stress-tested by nested sampling, the curved branches have not. Since the *priors* are wide enough to fully encompass the *posteriors*, there would be little quantitative effect on the conclusions of this paper from using a wider set of *priors*.

All *inference* products required to compute results are available for download from Zenodo[62].

[62] Handley, W. (August, 2019). Curvature tension: evidence for a closed universe (supplementary inference products). https://doi.org/ 10.5281/zenodo.3371152.

RESULTS

Tension results are shown in Fig. 2. For (ΛCDM), *Planck, lensing* and *Baryon acoustic oscillations* (BAO) are all consistent, with tensions ≤ 2. For KΛCDM, whilst BAO and *lensing* are consistent with one another, they are inconsistent with *Planck* at 2.5 σ, a tension similar to that between *Planck* 2015 and *weak galaxy lensing* (*Dark Energy Survey*). The *Hubble tension* between SH$_0$ES and *Planck* measured using the suspiciousness statistic is found to be $\sim$ 4.5 σ. Adding curvature generally enhances all tensions considered, except for *Planck+lensing* vs SH$_0$ES, for which the tension is weakened due to the increased error bar on H$_0$ from the inclusion of curvature.

Model comparison results are shown in Fig. 3. Without *lensing*, *Planck* prefers *curved universes* at a ratio of 50:1 with Ω_K = $-$ 4.5% $\pm$ 1.5%. Examining the *posterior* in Fig. 1, only 1/2000 of the *posterior mass* has $\Omega_K > 0$ *indicating a strong preference for closed universes over open ones.* When the moderately inconsistent *lensing data* are added, KΛCDM is only very slightly preferred to the *concordance model* at 2:1, but still prefers *closed universes* with Ω_K = $-$ 1.2% $\pm$ 0.6%. If one discounts the strong *discordance* of *Baryon acoustic oscillations* (BAO) and includes it in model constraints, *flat universes* become preferred in a Bayesian sense due to the effect of the Occam penalty penalizing the additional constrained parameter consistent with $\Omega_K = 0.0\% \pm 0.2\%$, *degenerate* with a *flat universe*.

The marginal distribution for the Planck, *lensing* and BAO *likelihoods* for all cosmological parameters are shown in Fig. 4.

CONCLUSIONS

In light of the inconsistency between *Planck*, CMB *lensing* and *Baryon acoustic oscillations* (BAO) data in the context of *curved universes*, *cosmologists can no longer conclude that observations support a flat universe*. If one assumes *Planck* CMB data are correct, then *we should conclude with odds of 50:1 that the universe is closed*.

These results by no means prove that the universe is *curved*. As with the *Hubble tension*, it may be that there are unaccounted systematic errors, bugs in some or all of the curved likelihood codes, or the possibility of *new physics* that could eventually release the parameter tension in K. What one can unambiguously say is that *further research is required to uncover why the cosmic microwave background* (CMB) *alone so strongly prefers a closed universe*, whilst other datasets provide quantitatively contradictory constraints.

The observational probes and codes used in this work are all nominally curvature-agnostic. However, since cosmological software packages and data compression mechanisms often require approximations, ducial biases toward *flatness* can easily creep in. A thorough investigation of these implicit assumptions forms the natural follow-up to this work. *Closed universe models* can generally relax the *Hubble tension* between *supernovae observations* and the *cosmic microwave background* (CMB). *If the solution to curvature tension resides in new physics, then it is possible that an additional change alongside the introduction of curvature may yet resolve both the Hubble and the curvature inconsistencies*. The fact that *curved cosmologies* relax the strong CMB constraint on the *Hubble constant* may aid additional modifications in finding a resolution to the *Hubble tension*.

Abdalla, E., *et al.* **(June, 2022). Cosmology intertwined: A review of the particle physics, astrophysics, and cosmology associated with the cosmological tensions and anomalies.**

Journal of High Energy Astrophysics, 34, 49-211; https://doi.org/10.1016/j.jheap.2022.04.002.

Abstract

The standard Λ Cold Dark Matter (ΛCDM) cosmological model provides a good description of a wide range of astrophysical and cosmological data. However, there are a few big open questions that make the standard model look like an approximation to a more realistic scenario yet to be found. In this paper, we list a few important goals that need to be addressed in the next decade, taking into account the current discordances between the different cosmological probes, such as the *disagreement in the value of the Hubble constant* H_0, *the* $\sigma_8 - S_8$ *tension*, and other less statistically significant anomalies. While these discordances can still be in part the result of systematic errors, their persistence after several years of accurate analysis strongly hints at cracks in the standard cosmological scenario and the necessity for new physics or generalizations beyond the standard model. *In this paper, we focus on the tension between the Planck CMB estimate of the Hubble constant and the SH0ES collaboration measurements.* After showing the evaluations made from different teams using different methods and geometric calibrations, we list a few interesting new physics models that could alleviate this tension and discuss how the next decade's experiments will be crucial. Moreover, we focus on the tension of the *Planck* CMB data with weak lensing measurements and redshift surveys, about the value of the matter energy density Ω_m, and the amplitude or rate of the growth of structure (σ_8, $f\sigma_8$). We list a few interesting models proposed for alleviating this tension, and we discuss the importance of trying to fit a full array of data with a single model and not just one parameter at a time. Additionally, we present a wide range of other less discussed anomalies at a statistical significance level lower than the $\sigma_8 - S_8$ tensions which may also constitute hints towards *new physics*, and we discuss possible generic theoretical approaches that can collectively explain the non-standard nature of these signals. Finally, we give an overview of upgraded experiments and next-generation space missions and facilities on Earth that will be of crucial importance to address all these open questions.

Lerner, E. J. (October, 2022)*. The Big Bang Never Happened—A Reassessment of the Galactic Origin of Light Elements (GOLE) Hypothesis and its Implications.

https://www.lppfusion.com/storage/GOLE-Lerner.pdf.

* Online paper updated to October, 2022. (63 pages.)

This work was funded by LPPFusion, Inc., 128 Lincoln Blvd., Middlesex, NJ.

> *LPPFusion website*:
> "These are the papers that the cosmology censors don't want anyone to read:
>> 1. Will LCDM cosmology survive the James Webb Space Telescope? By Riccardo Scarpa and Eric J. Lerner.
>> 2. Observations of Large-Scale Structures Contradict the Predictions of the Big Bang Hypothesis But Confirm Plasma Theory by Eric J. Lerner.
>> 3. The Big Bang Never Happened—A Reassessment of the Galactic Origin of Light Elements (GOLE) Hypothesis and its Implications by Eric J. Lerner.

The *first one* predicts what the new JWST telescope will find—further refuting the *Big Bang*, expanding universe, hypothesis. The *second paper* shows, with the latest data, how large-scale structures could not have formed in the time since the hypothesized *Big Bang*—and how they really formed from *plasma filamentation*. The *third paper* summarizes the evidence against the *Big Bang* hypothesis, which is contradicted by at least 16 independent sets of data and supported by only one. It also shows how a universe without a *Big Bang* evolved into the one that we currently observe.

These papers were refused publication even on the arXiv pre-print website that supposedly allows all researchers to publish without peer review.

Abstract

The growing list of failed predictions of the *inflationary LCDM theory* is a widely-recognized crisis in cosmology. It is therefore timely to re-examine if the *Big Bang hypothesis* (BBH), which underlies the dominant cosmological model, is valid. The core of that hypothesis is that the universe began with a short period of extremely high temperature and density. Such a hot, dense epoch produces light elements by fusion reactions. But the actual published predictions of the *Big Bang Nucleosynthesis* (BBN) theory of light element production have increasingly diverged from *observations*. The predictions for both *lithium and helium abundance* now differ by many standard deviations from *observations*, a situation that is worsening at an accelerating pace. Only *deuterium* predictions have remained in agreement with *observation*. In contrast, the published predictions of *the alternative hypothesis, that all light elements were created by thermonuclear and cosmic ray processes in young galaxies, have been repeatedly confirmed by observations.*

This paper reassesses the galactic origin of light element (GOLE) hypothesis in light of new calculations and recent observations. The GOLE predictions remain in good agreement with all relevant elemental abundance data sets and are contradicted by none. As well, *the expansion of space required by the Big Bang hypothesis* (BBH) *is directly contradicted by both data on surface brightness and supernova light curves. Nor are any of the quantitative predictions of BBH for the CMB in accord with observations*, while *Galactic Origin of Light Elements* (GOLE) hypothesis provides an alternative explanation for the CMB that requires none of the BBH's hypothetical entities, such as *dark matter* or *dark energy*. BBH predictions are contradicted by 16 different data sets while GOLE predictions are contradicted by none. *The solution to the crisis in cosmology is to abandon the Big Bang hypothesis.*

---

1. Introduction: Was there a hot dense Big Bang?

A key hypothesis of *concordance cosmology* is that the universe began with a brief epoch of *extremely high temperature and density*, the *Big Bang*. The most direct test of the validity of this theory is in comparing its predictions with the abundance of certain light elements, since if the universe went through a brief period of high density and temperature, *thermonuclear reactions* would inevitably produce *helium, deuterium* and *lithium* in specific amounts that depend only on the ratio of *photons* to *baryons*. It has been repeatedly claimed by many authors that the predictions of the theory have been confirmed by observation and thus that the hypothesis is valid (for example, Mathews, Kusakabe, & Kajino, 2017)[1].

> [1] Mathews, G. J., Kusakabe, M., & Kajino, T. (June, 2017). Introduction to Big Bang Nucleosynthesis and Modern Cosmology. *International Journal of Modern Physics*, 26, 08, 1741003; https://doi.org/10.1142/S0218301317410014.

But an objective assessment of the abundant literature leads to the opposite conclusion, that observations contradict the predictions and that improved observations have increased this contradiction, invalidating the hypothesis. Since this conclusion is opposed to many published claims, *this paper presents in section 2 an in-depth analysis of this divergence of predictions and observations.* We then, *in section 3, show that the GOLE hypothesis has correctly predicted observations* and *in section 4 that the agreement still exists and has improved with time.* Since the *Big Bang hypothesis* (BBH) is widely claimed to be indirectly supported by many other data sets, including *evidence of the expansion of the universe*, a full test of its validity must examine these claims, as well as testing if the GOLE hypothesis is compatible with these other data sets. *The implications of the GOLE and BBH for the antimatter problem* are examined in section 5; for the *expansion of the universe* and the *surface brightness test* in section 6; and for *supernova light curves* in section 7. *The results are discussed in section 8 and conclusions arrived at in section 9. ...*

8. The Cosmic Microwave Background Radiation

A fourth and final major difference in implications between *Big Bang hypothesis* (BBH) and GOLE concerns the *cosmic microwave background* (CMB) radiation. For the BBH, the CMB is the extremely *redshifted* radiation from the initial hot, dense *Big Bang* epoch. However, the *cosmic microwave background* (CMB), which is highly thermalized and highly isotropic, does not constitute direct evidence of such an epoch. Rather it is direct evidence that, at some point in cosmic history, *the universe must have had a high optical depth for the wavelength band covered by the CMB.*

> [*Optical depth* is the natural logarithm of the ratio of *incident* to *transmitted* radiant power through a material. Thus, *the larger the optical depth, the smaller the amount of transmitted radiant power through the material.*]

Observations show that the *CMB spectrum* follows a *black body spectrum* to within about 5×10^{-5} in the *frequency* range from 40-200 GHz (wavelength band 1.5-7.5 mm) and to within 5×10^{-4} in a broader range from 20-400 GHz (wavelength band 0.75-15 mm).

> [The *frequency* of *cosmic microwave background* (CMB) *radiation* ranges from 0.3 GHz to 630 GHz.
>
> The radio frequencies from 0.3 GHz to 300 GHz are referred to as *microwaves* (wavelength from about one meter to one millimeter). *Microwaves* travel by line-of-sight; unlike lower frequency radio waves, they do not diffract around hills, follow the earth's surface as ground waves, or reflect from the ionosphere, so terrestrial *microwave* communication links are limited by the visual horizon to about 64 km. At the high end of the band, they are absorbed by gases in the atmosphere, limiting practical communication distances to around a kilometer. The frequencies above 300 GHz (wavelengths above one meter) are in the *far-infrared* (which ranges from 300 GHz to 30 THz) The

lower part of this range may also be called microwaves or terahertz waves. This radiation is typically absorbed by so-called rotational modes in gas-phase molecules, by molecular motions in liquids, and by phonons in solids.

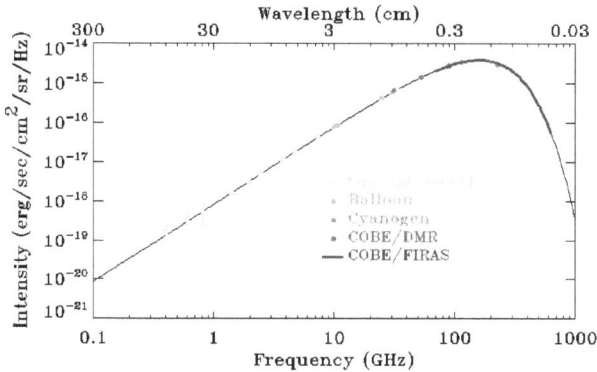

Spectral intensity is the *radiant flux* emitted, reflected, transmitted, or received *per unit solid angle* and *per unit frequency* or wavelength.

[This graph *redrawn against a linear frequency scale* appears as an almost horizontal block at an *intensity* rising sharply from 10^{-20} to 10^{-15} erg^{-1} cm^{-2} sr^{-1} at around 10 GHz, then rising slowly to a peak of about 5 x 10^{-15} erg^{-1} cm^{-2} sr^{-1} at around 110 GHz, then falling back at around 630 GHz, where sr is a steradian, which is the *unit of solid angle* measure (there are 4π steradians in a complete sphere).]

The *cosmic microwave background* (CMB) has been used as landmark evidence of the *Big Bang theory* for the origin of the universe. In the *Big Bang* cosmological models, during the earliest periods, the universe was filled with an opaque fog of dense, hot *plasma* of sub-atomic particles. As the universe expanded, this *plasma* cooled to the point where *protons* and *electrons* combined to form neutral atoms of mostly *hydrogen*. Unlike the *plasma*, these atoms could not scatter thermal radiation by *Thomson scattering*, and so the universe became transparent. Known as the *recombination epoch*, this *decoupling* event released *photons* to travel freely through space. However, *according to this version of the Big Bang theory*, the photons have grown less energetic due to the cosmological redshift associated with the expansion of the universe. (This is a reversal of the previous argument that *the cosmological redshift was due to the expansion of space*, not due to the *photons* losing energy as they travel through intergalactic space.)]

The *cosmic microwave background* (CMB) is also *isotropic* to with a few parts in 10^5. Since the deviation of a spectrum from a *blackbody* depends on *optical depth*

$$F(v) = 1 - e^{-t(v)}, \tag{12}$$

where $F(v)$ is the ratio of *observed flux* at a given *frequency* to *blackbody flux* and t (v) is the *optical depth* at that *frequency*, then the CMB observations directly imply an *optical depth* of $\ln(2 \times 10^4) = 10$ for the narrower *frequency* band and $\ln(2 \times 10^3) = 7.6$ for the broader band. The CMB observations thus demonstrate that the average *photon* in these bands has suffered 7.6-10 *large angle scatterings* or, equivalently, the *same number of absorptions and re-emissions*, from initial *emission* to *observation*.

The question then is: what conditions produced such a *high optical depth* in these frequency bands? For the *Big Bang hypothesis* (BBH), the conditions were a *hot dense plasma* at a temperature of about 10^4 K, producing a *black body* spectrum *that was then redshifted a few thousand-fold to the present bands*. But for three decades, *there has been an alternative explanation for the cosmic microwave background* (CMB). That is that the energy for the CMB derives from the *thermonuclear reactions* in *stars* that produced the observed abundance of He, and that the radiation is thermalized and isotropized by the *intergalactic medium*, which has a *high optical depth* in the present frequency bands of the CMB … .

> [There is no doubt that this occurs. It is possible that both contributed to the *cosmic microwave background* (CMB), but, as in the case of the cause of the *cosmological redshift*, if the *thermonuclear reactions* in *stars* were sufficient, then this suggest that *there was no Big Bang.* See Underwood, T. G., (2024). *Cosmological Redshift of Light*, p. 11.
>
> The articles in this volume describing the alternative sources of foreground *anomalous microwave emission* (AME) demonstrate the processes by which radiation from *thermonuclear reactions* in *stars* can be thermalized by atoms or dust in the *intergalactic medium*, explaining both the *cosmological redshift* and the *cosmic microwave background* (CMB).]

Despite the widespread popularity of the BBH explanation of the CMB, the test of its validity remains its ability to correctly predict observations. *In the past five years, and especially in the past two years, multiple predictions of the BBH theory of the CMB have been contradicted by independent data sets.* The failure of the predictions of the theory for H_0, in particular, have led to a growing awareness of a crisis in cosmology.

8.1 Quantitative Predictions of BBH Theory of CMB vs Observations

In assessing any theory, it is crucial to compare predictions with subsequent *observations*. In the first 30 years after the discovery of the *cosmic microwave background* (CMB), the quantitative predictions of *the BBH theory of the CMB were consistently contradicted by subsequent observations*, which was then followed by the addition of ad-hoc hypotheses to the theory to fit already-observed data. The BBH hypothesis in its original form *did not predict a smooth isotropic CMB* because there was no mechanism allowing distant parts of the sky to reach equilibrium with each other during the *Big Bang* expansion. *This required the introduction of the inflation hypothesis in 1980, an entirely ad-hoc hypothesis introducing a new inflation field that is not otherwise observed.* Second, the original predictions for the amplitude of small fluctuations in the CMB, about 10^{-4} to 10^{-3}, turned out to be far too large compared with observations of fluctuations of at most 10^{-5}. *This observation required the introduction of the hypothesis of non-baryonic or dark matter (DM) in order to produce large-scale structure.*

However, the *inflation hypothesis* predicted a *flat universe* with $\Omega_{tot} = 1$ and thus predicted a *total dark matter density* that was also close to $\Omega_{tot} = 1$. By the mid-1990's the *Cold Dark Matter* (CDM) model also encountered gross contradictions with observations. In particular, as measurements of the *Hubble relation* improved, this model predicted an age for the universe of about 8 Gy (Krisciunas, 1993[1]).

[1] Krisciunas, K., (August, 1993). Look-Back Time, the Age of the Universe, and the Case for a Positive Cosmological Constant. *J. Royal Astr. Soc. Can.*, 87, 4, 223; https://adsabs.harvard.edu/full/1993jrasc..87..223k.

This was far too short, as evidence from both individual stars and galaxies indicated. As well, the predicted deviation from linearity of the *Hubble relation* for *supernovae* was not observed. *This led to the ad-hoc introduction of yet a third hypothesis, dark energy,* (also termed the *cosmological constant*) symbolized by Λ, *thus producing the current ΛCDM model* (Primack, 1995[2]).

[2] Primack, R., (1995). Status of Cosmological Parameters: Can $\Omega=1$? arxiv/9503020; https://doi.org/10.48550/arXiv.astro-ph/9503020.

This complex model, despite its many ad-hoc additions, retains testable predictions. Inflation by itself predicts that the fluctuations in the *cosmic microwave background* (CMB) should be isotropic and Gaussian—a random pattern—and that the geometry of the universe should be flat (Schwarz, *et al*, 2016[3]).

[3] Schwarz D. J., Copi, C. J., Huterer, D., & Starkman, G. D. (August, 2016). CMB anomalies after Planck. *Class. Quantum Grav.*, 33, 184001; https://doi.org/10.1088/0264-9381/33/18/184001. See above.

When further complicated by additional hypotheses, the ΛCDM also predicts the exact amplitude of the fluctuations' spatial spectrum *with the adjustment of at least 8 free parameters*: the *mass density of baryonic matter, dark matter, neutrinos and dark energy*, the *optical depth* at the time of reionization, the "*bias factor*" (relating the clumping of dark matter and baryonic matter), the *number of neutrino species* and the *Hubble constant*. This list, however, assumes the *flatness of space*. If that prediction is also to be tested, there are (at least) 9 free parameters to the model.

In the late 1990's this new model did begin to claim accord with subsequent observations. In 1996, analysis of COBE *observations* indicated that the *anisotropies* in the CMB were Gaussian at all scales, as predicted by inflation (Primack, 1995). By 2003, based on early results from WMAP, published fits of the power spectrum of CMB *anisotropies* claimed close agreement with the H_0 and *total matter density* obtained by independent fits to the SN1a *luminosity-distance* correlations (Hinshaw, *et al.*, 2003[4]).

[4] Hinshaw, G., *et al.* (2003). First-Year *Wilkinson Microwave Anisotropy Probe* (*WMAP*). Observations: The Angular Power Spectrum. *Astrophys. J. Supp.*, 148, 135; https://doi.org/10.1086/377226.

In 2006 additional data analysis (Sanchez, *et al.*, 2006[5]) showed a fit to the *inflation* prediction of a *flat universe* and good correspondence to the power spectrum all the way to the largest modes, although the largest, *quadrupole mode* seemed surprisingly small.

[5] Sánchez, A. G., Baugh, C. M., Percival, W. J., Peacock, J. A., Padilla, N. D., Cole, S & Baugh, C. M. (February, 2006). Cosmological parameters from cosmic microwave background measurements and

the final 2dF Galaxy Redshift Survey power spectrum. *MNRAS*, 366, 1, 189-207; https://doi.org/10.1111/j.1365-2966.2005.09833.x.

Thus by 2006, it appeared that the ΛCDM variant of *Big Bang hypothesis* (BBH), with its 8 free parameters, had made at least five independent quantitative predictions that were in accord with subsequent observations. However, there were already clouds on the horizon. As early as 2004, a number of analyses (Schwarz, *et al*, 2004[6]) started to point out significant deviations in the WMAP data from *Gaussianity*.

[6] Schwarz, D. J., Starkman, G. D., Huterer, D., & Copi, C. J. (November, 2004). Is the Low-ℓ Microwave Background Cosmic? *Phys. Rev. Lett.*, 93, 221301; https://doi.org/10.1103/PhysRevLett.93.22130.

Further releases of WMAP data, and especially the release of multi-year Planck data after 2013, confirmed and greatly strengthened these analyses so that by 2016 published analyses (Schwarz, *et al*, 2016[3]) showed that *the prediction of Gaussianity was clearly contradicted*.

These contradictions occur on the largest scales observable—in this case the angular scales larger than a few degrees ($l < 30$). All of the *cosmic microwave background* (CMB) contradictions have been discussed widely in the literature, by many authors, with Schwarz, *et al*, 2016[3] providing a convenient summary. *Isotropy is strongly contradicted* because the octopole mode is strongly planar, and because the quadrupole and octopole planes are aligned with each other and aligned with the CMB dipole, theoretically produced by the Solar Systems motion through the CMB (Fig.16). In addition, the *total amplitude of fluctuations* varies according to the hemisphere of the sky observed and the odd multipoles have different power than the even multipoles, particularly in the range $20 < l < 30$ (Fig. 17). A recent analysis (Yeung & Chuy, 2022[7]) shows that the calculated predictions of cosmological parameters from the Planck data shows strong *anisotropy* as well.

[7] Yeung, S. & Chuy, M. -C. (2022). *Phys Rev.*, D, 105, 083508.

The theory predicts strong correlation of the *cosmic microwave background* (CMB) at large angles, but *the correlations at angle > 70 degrees are far smaller than predicted* (Fig. 18). Each of these contradictions, which are independent of each other, have a probability to occur by chance of less than 0.5% (if *inflation* were a valid hypothesis) and therefore have negligible probability, less than one in 10^7, taken together.

By the same time, *Planck* data made clear that for the power spectrum, the best fit curve for $l > 30$ was a poor fit for $l < 30$. In particular, as can be seen if Figure 19, the strong dip in amplitude in the range $15 < l < 30$ is entirely missed in the predicted curve. This is the same range as for the greatest parity and hemispheric asymmetry. So *even with 9 free parameters, the entire data set cannot be accurately fit.* Thus by 2016, published analyses unequivocally demonstrated that *two basic quantitative predictions of the Big Bang hypothesis* (BBH) *theory of the cosmic microwave background* (CMB) *were contradicted by better data.*

In the past two years, *the three remaining quantitative predictions have been contradicted by data.* Handley (2019)[8], Park & Ratra (2019)[9] and Di Valentino, *et al*, (2020)[10] have shown that when the curvature parameter is allowed to vary, the best fit to the power spectrum is *not a flat universe*, contradicting another basic prediction of *inflation*.

[8] Handley, W. (February, 2021). Curvature tension: evidence for a closed universe. *Phys. Rev.*, D, 103, L041301; https://doi.org/ 10.1103/PhysRevD.103. L041301. See above.

[9] Park, C. -G., & Ratra, B. (December, 2019). Using the tilted flat-ΛCDM and the untilted non-flat ΛCDM inflation models to measure cosmological parameters from a compilation of observational data. *Astrophys. J.*, 882, 158; https://arxiv.org/pdf/1801.00213.

[10] Di Valentino, E., Melchiorri, A. & Silk, J. (February, 2020). Planck evidence for a closed Universe and a possible crisis for

cosmology. *Nat. Astron.*, 4, 196–203; https://doi.org/10.1038/s41550-019-0906-9.

It is important to emphasize. that without *inflation*, the BBH *does not even predict an approximately isotropic CMB*, so *is grossly contradicted* by the observations of fluctuations at the level of 10^{-5}.

The cosmic microwave background (CMB) considered by itself thus provides extremely serious contradictions to BBH predictions. There are as well additional serious contradictions when the BBH fit to the CMB power spectrum is used to make predictions about other data sets. In the past two years, much attention was given to the fact that the value of the *Hubble constant* predicted by the fit to the *CMB spectrum* is different from the value actually measured by comparing the distance to *supernovae* with their *redshifts*, with the difference now reaching 4-6 σ (Riess, 2020[11]).

[11] Riess, A. (January, 2020). The expansion of the Universe is faster than expected. *Nature Reviews Physics*, 2, 10-12; https://doi.org/10.1038/s42254-019-0137-0.

While this has been described as a difference between two measurements, *it is in fact another failure of a theoretical prediction, based on fitting the CMB spectrum with the help of Big Bang theory.* The only direct measurement was based on the *supernovae* data, which compare two observable quantities, the apparent brightness of the *supernovae* and the *redshift* of their spectrums.

It is important to note that the work over the past two years by Kang, *et al* (2020)[12] and others showing the correlation of *stellar population age* with *SN1a luminosity* has demonstrated that when this correlation is taken into account, the *SN1a* data are best fit (assuming the BBH) with $\Lambda = 0$.

[12] Kang, Y., Lee, Y. -W., Kim, Y. -L., *et al.* (2020). *Astrophys. J.*, 889, 8.

Since the *cosmic microwave background* (CMB) power spectrum is fit with $\Lambda = 0.7$, this demonstrated *yet other contradiction in the BBH predictions*.

Another extremely significant contradiction is in the prediction, based on CMB fitting and the ΛCDM model, for the *density of dark matter*. *There is convincing evidence from multiple data sets that non-baryonic matter does not exist at all* [Adhikari, G., *et al.* (2018)[13]; Aprile, E., *et al.* (XENON Collaboration), (2018)[14]; Kroupa, P., Pawlowski, & Milgrom, M., (2012)[15]; Müller, *et al.*, (2018)[16]].

[13] Adhikari, G., *et al.* (December, 2018). The COSINE-100 Collaboration. An experiment to search for dark-matter interactions using sodium iodide detectors. *Nature*, 564, 83-6; https://doi.org/10.1038/s41586-018-0739-1.

[14] Aprile, E., *et al.* (XENON Collaboration). (September, 2018). Dark Matter Search Results from a One Ton-Year Exposure of XENON1T. *Phys. Rev. Lett.*, 121, 111302; https://doi.org/10.1103/PhysRevLett. 121.111.

[15] Kroupa, P., Pawlowski, M., & Milgrom, M. (December, 2012). The failures of the standard model of cosmology require a new paradigm. *Int. J. Mod. Phys.*, D, 21, 1230003; https://arxiv.org/pdf/1301.3907.

[16] Müller, O., *et al.* (February, 2018). A whirling plane of satellite galaxies around Centaurus A challenges cold dark matter cosmology. *Science*, 359, 534-7; https://doi.org/10.1126/science.aao1858.

But the ΛCDM model, combined with the CMB fitting, also predicts the *total amount* of *gravitating matter*. In the *flat universe* fit, this total is close to $\Omega_m = 0.3$. If *flatness* is not assumed, the best fit predicts even more *matter*, around $\Omega_m = 0.5$ [Schwarz, *et al*, (2016)[3]]. But this prediction for *total matter density* is strongly contradicted by measurements that relate *galaxy cluster* distribution to *observed velocities* of *bulk flows* of *galaxies*. These studies [Karachentsev, (2012)[17]] show that even on large scales $\Omega_m < 0.1$, which is 10 σ from the *Big Bang hypothesis* (BBH) predictions.

[17] Karachentsev, I. D. (May, 2012). Missing dark matter in the local universe. *Astrophys. Bull.* 67, 123–34; https://doi.org/10.1134/S1990341312020010.

Thus, taking together the *large-scale asymmetries*, the *poor fit at large angles*, the *contradictions with flatness predictions*, the *wrong prediction of matter density* and the widely-discussed *wrong prediction of the Hubble relation, the Big Bang hypothesis (BBH) predictions of the CMB are amply falsified. At present, no quantitative predictions of the BBH for the CMB are in accord with observations.* As Di Valentino, *et al.* summarize: "BAO surveys disagree at more than 3 standard deviations. CMB lensing is in tension at the 95% CL. The R18 constraint on the Hubble constant is in tension with PL18 at more than 5 standard deviations, while cosmic shear data disagree at more than 3 standard deviations. These inconsistencies between disparate observed properties of the Universe introduce a problem for modern cosmology: the flat ΛCDM model, de facto, does not seem any longer to provide a good candidate for concordance cosmology…". [DiValentino, Melchiorri & Silk, (2020)[10]]. To this, we add only that the "*flat ΛCDM model*" is the current version of the BBH.

This is not the viewpoint of a small fraction of researchers in the field. A comprehensive survey of the "anomalies" between *concordance cosmology* and *observations* by some 200 authors [Abdalla, *et al*, (2022)[18]] listed at least 22 such contradictions.

[18] Abdalla, E., *et al.* (June, 2022). Cosmology intertwined: A review of the particle physics, astrophysics, and cosmology associated with the cosmological tensions and anomalies. *Journal of High Energy Astrophysics*, 34, 49; https://doi.org/10.1016/j.jheap.2022.04.002.

8.2 The non-expanding (GOLE) theory of the CMB

There is an alternative explanation of the *Cosmic Microwave Background* (CMB) that does not require a *Big Bang*. As has been noted repeatedly in the past and is pointed out again in section 4.2, *the energy needed to account for the microwave background is comparable to the energy that would have been released by the production by ordinary stars of the known amount of helium,* as predicted by the *Galactic Origin of Light Elements* (GOLE) hypothesis.

Calculations in section 4.2 and earlier papers predicted a *thermonuclear energy* release of 4 Mev/H. Given this value, the known CMB *energy density* of 0.26eV/cm^3 would be produced from a total H *density* of 6.6 x 10^{-8}/cm^3. By comparison, the *baryonic density* in the current Big Bang model is 2.5 x 10^{-7}/cm^3 while actual observations of *hydrogen density* on large scales are around 1.2 x 10^{-7}/cm^3. Using this latter *observational* value predicts an *energy release* that is greater than that needed to produce the *observed* 2.73 K *background radiation temperature*, if it is thermalized to equilibrium.

The *isotropy* and *black-body* spectrum of the *cosmic microwave background* (CMB) are inevitable if the present-day *intergalactic medium* has sufficient *optical depth* in the *microwave band*. The present author proposed (Lerner, 1988[19]) along with others (Peter & Peratt, 1990[20]) that the CMB is a *radio fog* permeating the present-day universe, not some ghost of a long-ago *Big Bang*.

[19] Lerner, E. J. (August, 1988). Plasma model of microwave background and primordial elements: an alternative to the big bang. *Laser and Particle Beams*, 6, 456-69; https://doi.org/10.1017/S0263034600005395.

[20] Peter, W., & Peratt, A. L. (1990). *IEEE Trans. Plasma Sci.*, 18, 49.

As Planck demonstrated in deriving the *blackbody* spectrum, any body with high *optical depth* must radiate with a *blackbody* spectrum.

There is a clear-cut test for the hypotheses that there is a significant optical depth in the present-day universe. If the universe *scatters or absorbs microwave and radio radiation,* but is transparent in the shorter-wavelength infrared bands, then more distant objects will appear dimmer in radio bands then in IR. *Observational evidence* (Lerner,

1990[21], 1993[22]) of this absorption effect comparing 60μ - 100μ IR with 1.4GhZ radio showed that radio emission by galaxies dropped by a factor of 10 as distance increased to 300 Mpc.

[21] Lerner, E. J. (September, 1990). Radio absorption by the intergalactic medium. *Astrophys. J.*, 361, 63-8; https://adsabs.harvard.edu/full/1990ApJ...361...63L.

[22] Lerner, E. J. (September, 1993). Confirmation of radio absorption by the intergalactic medium. *Astrophys. & Space Sci.*, 207, 17-26; https://doi.org/10.1007/BF00659126.

This work showed that for nearby galaxies with redshift z < 0.07, radio emission for a given IR luminosity falls as $z^{-0.32}$. (Fig. 20) A zero correlation with z, as would be expected in an *intergalactic media* (IGM) fully transparent to radio radiation, was excluded at a 5 σ level in (Lerner, 1990)[21] and at an 8 σ level in (Lerner, 1993)[22].

This distance of 300 Mpc corresponds to a look-back time of only 900 My, too short a time for any evolutionary process to create such a dramatic change. An evolutionary process accounting for this change in brightness with distance would have to accelerate unphysically as it approached the present, with galaxies doubling in brightness in the last 10 million years alone.

This strong decrease in radio emission with z was overlooked by many other researchers, because the widespread measure of the radio-IR relationship is q, the log ratio of IR to radio emission. This measure assumes a linear relationship between the two *luminosities*. However, it has long been clear that the actual relationship is non-linear. As Devereux & Eales (1989)[23] first found and later work (Lerner, 1990)[21] confirmed, the $L_r \sim L_{IR}$ 1.29.

[23] Deveraux, N. A., & Eales, S. A. (May, 1989). A reevaluation of the infrared-radio correlation for spiral galaxies. *Astrophys. J.*, 340, 708; https://doi.org/10.1086/167431.

Since mean L_{IR} is steeply correlated with z, therefore at higher z, more luminous galaxies are observed, with higher radio-IR ratios (lower q). This effect obscures the fall with z in *radio luminosity* for the same IR *luminosity*. Simply analyzing the data among the three variables simultaneously yields the nonlinear *radio-IR relationship* and the steep decrease in *radio luminosity* with z.

This evidence of *radio luminosity* decrease is here further extended to 600 Mpc with new data at 150 MHz. Using LOFAR and IRAS data, Wang, *et al.* (2019)[24] found that

$$L_{150} \sim L_{IR}{}^{1.37+-0.045}. \tag{13}$$

[24] Wang, L., *et al.* (2019). *Astronomy and Astrophysics*, 631, A109.

The re-analysis here of Wang, *et al.*'s data presented here uses the 412 galaxies identified as star forming or starburst, excluding broadline galaxies and AGNs. This shows a dependency of $L_{150} \sim L_{IR}{}^{1.29+-0.048}$, in statistical agreement with Wang's and with the 1.29 slope found previously by Lerner and Devereux & Eales for 1.4 GHz (fig. 21).
A linear slope is ruled out at a 6 σ level. The same analysis shows that the *radio luminosity* for a given L_{IR} declines as $z^{-0.42\pm0.087}$, (Fig.20 b) again in good agreement with the exponent of -0.41 ± 0.06 in (Lerner, 1990)[21] and -0.32 ± 0.04 in (Lerner, 1993[22]). The zero slope expected with a transparent *intergalactic media* (IGM) is ruled out again at a 4.8 σ level.

This is an example of an assumption, made in support of the *Big Bang hypothesis* (BBH), of a relationship among multiple variables that is not supported by data. A second one is the assumption of a *flat universe* in the analysis of the *cosmic microwave background (CMB)*. Eliminating these false assumptions, and using instead the correlations derived from the actual data, makes clear the contradictions between the BBH predictions and the data.

At higher *redshifts*, present data makes measurements of *radio luminosity* more model-dependent, since they depend on K corrections at both IR and radio *wavelengths*, which are uncertain, and any changes in the *radio-IR relationship* could be attributed to evolution. However, spectra of *ultraluminous infrared galaxies* (ULIRGs) at high z (fig. 22) show a reduced ratio of FIR to NIR radiation (Sajina, *et al.*, 2012[25]) as would be expected if *absorption* or scattering occurs at *wavelengths* longer than about 200 μ.

[25] Sajina A., Yan, L., Fadda, D., Dasyra, K., & Huynh, M. (September, 2012). Spitzer- and herschel-based spectral energy distributions of 24 μm bright z ∼ 0.3-3.0 starbursts and obscured quasars. *Astrophys. J.*, 757, 13; https://doi.org/10.1088/0004-637X/757/1/13.

This effect also could be attributed to evolution, but is entirely consistent with the predictions of an *opacity model*.

The data presented in fig. 20, while limited in z, can be used to estimate a range of *optical depths* for the *intergalactic media* (IGM). It is important to note that the data itself, with no extrapolation, already shows an absorption of a factor of ten, and thus an *optical depth* $\tau = 2.3$, 30% of that needed to account for the *CMB spectrum* in the frequency range up to about 20 GHz. We also note that the fit to a power law dependency on distance is consistent with a fractal distribution of *scatterers* or *absorbing objects* with fractal dimension of 2. Such a density distribution would reflect the observed fractal distribution of *galaxies* (Telesa, *et al.*, 2021[26]).

[26] Telesa, S., Lopesb, A. R., & Ribeiroa, M. B. (February, 2021). Fractal analysis of the UltraVISTA galaxy survey. *Phys. Lett.*, B, 813, 136034; https://doi.org/10.1016/j.physletb.2020.136034.

Thus, if density $n = k/D$, then the optical depth $\tau = k \ln D/Mpc - \ln D_0/Mpc$, where D_0 is the *distance* where the fractal rise in *density* plateaus. The observed *luminosity* after *absorption* or *scattering* then is proportional to $e^{-\tau}$ or D^{-k}, thus producing the observed power law.

The *absorption* can be extrapolated to large distance if we hypothesize at what *distance*, D_H, the fractal distribution of *density* levels out into a homogenous distribution. We take the minimum D_H to be just the maximum *distance* already observed in the data—600 Mpc. The maximum D_H is equal to the 1.5 Gpc *radius* of the largest inhomogeneities observed at present (Shirokov, *et al.*, 2016[27]).

[27] Shirokov, S. I., Lovyagin, N. Y., Baryshev, Y. V., *et al.* (June, 2016). Large-scale fluctuations in the number density of galaxies in independent surveys of deep fields. *Astron. Rep.*, 60, 563–78; https://doi.org/10.1134/S1063772916040107.

We take D_0 to be 1 Mpc as a maximum, since the data shown in fig. 20a shows no sign of a plateau at larger D.

We are assuming a *non-expanding, non-BBH universe* as described by section 6, with D = cz/H_0. With this assumption and the hypothesized range for D_H we then can calculate that an optical $\tau = 7.6$ will result for *absorption* up to a z = 2.2 ± 1.0 and $\tau = 10$ will result for a z = 4.3 ± 2. *It is thus clear that extrapolating the observed radio absorption produces the right order of magnitude of τ needed to produce the observed cosmic microwave background* (CMB) *blackbody spectrum and isotropy.*

This data is limited to frequencies well below those in the frequency range 20-400 GHz where the *cosmic microwave background* (CMB) is closest to a *blackbody*. Future work will look at higher frequency data. However, there is no evidence that the *optical depth* of the *intergalactic media* (IGM) decreases sharply with increasing frequency above 1 GHz. If this were the case, localized sources at substantial z would show a kink in the spectrum, with the slope α of I_ν on ν decreasing in the region of $\nu > 1$ GHz. But this is not observed. For example, a sample of sources selected at 15.7 GHz show a median α of -0.6 (Whitman, *et al.*, 2015), essentially the same α as for the MW and other galaxies measured in the range $\nu < 1$ GHz.

8.3 Mechanisms for absorption or scattering

There are at least two phenomena that could account for the *absorption* of *microwave* and *RF* radiation. One is that radiation in these *wavelengths* could be *absorbed* and reemitted by *electrons* trapped in dense *plasma filaments* emitted from a range of *astrophysical jets* extending from *stellar Herbig-Haro objects* to *quasars* (Lerner, 1993)[22].

[*Plasma* is one of four fundamental states of *matter* (the other three being *solid, liquid,* and *gas*) characterized by the *presence of a significant portion of charged particles in any combination of ions or electrons*. It is the most abundant form of ordinary *matter* in the universe, mostly in *stars* (including the Sun), but also dominating the rarefied *intracluster medium* and *intergalactic medium*. *Plasma* can be artificially generated, for example, by heating a neutral gas or subjecting it to a strong electromagnetic field. The presence of *charged particles* makes *plasma electrically conductive*, with the dynamics of individual particles and macroscopic *plasma* motion governed by collective *electromagnetic fields* and very sensitive to externally applied fields.

Galaxy filaments are the largest known structures in the universe, consisting of walls of *galactic superclusters*. These massive, thread-like formations can commonly reach 50/h to 80/h megaparsecs (160 to 260 megalight-years)—with the largest found to date being the Hercules-Corona Borealis Great Wall at around 3 gigaparsecs (9.8 Gly) in length—and form the boundaries between voids. *Galaxy filaments* form the cosmic web and define the overall structure of the observable universe.

Herbig–Haro (HH) objects are bright patches of nebulosity associated with newborn stars. They are formed when narrow jets of partially ionized gas ejected by stars collide with *nearby clouds of gas and dust* at several hundred kilometers per second. *Herbig–Haro objects* are commonly found in star-forming regions, and several are often seen around a single star, aligned with its rotational axis. Most of them lie within about one parsec (3.26 light-years) of the source, although some have been observed several parsecs away. *Herbig–Haro objects* are transient phenomena that last around a few tens of thousands of years. They can change visibly over timescales of a few years as they move rapidly away from their parent star into the gas clouds of *interstellar space* (the *interstellar medium* or ISM).]

> [*According to the Big Bang theory*, due to the accelerating expansion of the universe, the individual clusters of *gravitationally bound galaxies* that make up *galaxy filaments* are moving away from each other at an accelerated rate; in the far future they will dissolve.]

In addition, *other researchers have pointed out that spinning dust particles can also absorb and re-emit microwaves* (Draine & Lazarian, 1998, see above). We do not consider this hypothesis further here.

The discovery (Leitch, et al., 1997[28]) of anomalous microwave emission (AME) from clouds in the Milky Way confirms that there exist processes that efficiently emit, and therefore can absorb and scatter, radiation in the cosmic microwave background (CMB) wavebands.

[28] Leitch, E. M., Readhead, A. C. S., Pearson T. J., & Myers, S. T. (September, 1997). An Anomalous Component of Galactic Emission. *Astrophys. J.*, 486, L23-6; https://doi.org/10.1086/310823.

It is important to note that a *blackbody* spectrum results from *any collection of absorbers that are sufficiently opaque*. Also, *scattering at large angle, where the emission angle is uncorrelated with the absorption angle, does not result in blurring of distant objects*, as small-angle scattering does.

Recently, evidence has been published of the existence of *plasma filaments* that are compatible with the *filaments* were predicted in

Lerner, 1992[29], 1993[22] to be capable of producing the *thermalization* of the *cosmic microwave background* (CMB). In these papers, the present author demonstrated analytically that *plasma filaments* would form from the beams emitted by QSO's and *active galactic nuclei* (AGN).

[29] Lerner, E. J. (December, 1992). Force-Free Magnetic Filaments and the Cosmic Background Radiation. *IEEE Trans. Plasma Sci.*, 20, 6, 935-8; https://ieeexplore.ieee.org/document/199554.

[An *active galactic nucleus* (AGN) is a compact region at the center of a *galaxy* that emits a significant amount of *energy* across the electromagnetic spectrum, with characteristics indicating that this *luminosity* is not produced by the *stars*. Such excess, non-stellar emissions have been observed in the radio, microwave, infrared, optical, ultra-violet, X-ray and gamma ray wavebands. A *galaxy* hosting an AGN is called an *active galaxy*. The non-stellar radiation from an AGN is theorized to result from the accretion of *matter* by a supermassive *black hole* at the center of its host *galaxy*.

A *quasar* is an extremely luminous *active galactic nucleus* (AGN). It is sometimes known as a *quasi-stellar object* (QSO). The emission from an AGN is believed to be powered by accretion onto a supermassive *black hole* with a *mass* ranging from millions to tens of billions of solar masses, surrounded by a gaseous accretion disc. Gas in the disc falling towards the black hole heats up and releases energy in the form of *electromagnetic radiation*. The radiant energy of *quasars* is enormous; the most powerful *quasars* have *luminosities* thousands of times greater than that of a *galaxy* such as the Milky Way. The *redshifts* of *quasars* are of cosmological origin.

The term *quasar* originated as a contraction of "quasi-stellar [star-like] radio source"—because they were first identified during the 1950s as sources of radio-wave emission of unknown physical origin—and when identified in photographic images at visible wavelengths, they resembled faint, star-like points of light. High-resolution images of *quasars*, particularly from the *Hubble Space Telescope*, have shown that *quasars* occur in the centers of *galaxies*, and that some host *galaxies* are strongly interacting or merging *galaxies*. As with other categories of AGN, the observed properties of a *quasar* depend on many factors, including the *mass* of the *black hole*, the rate of *gas accretion*, the *orientation of the accretion disc relative to the observer*, the *presence or absence of a jet*, and the *degree of obscuration by gas and dust* within the host *galaxy*.]

For clarity, we here summarize the much more detailed derivations in those papers. Gaussian units are used.

We assume that in the accelerating region of a QSO or AGN *ion* beam that *ions* and *electrons* travel along *field lines*, that *magnetic field energy* is approximately equal to *ion kinetic energy* and that only *current* carriers are present in the beam (as background particles would be rapidly expelled). We also assume for simplicity a pure H *plasma*. We then have:

$$B^2/8\pi = n_i\gamma_pMc^2 \tag{14}$$
$$n_i = B/2\pi er = B^2/8\pi\gamma_pMc^2 \tag{15}$$
$$B = 4\gamma_pMc^2/er \tag{16}$$
$$I = 2\gamma_pMc^3/e = 2\gamma_pec/r_p, \tag{17}$$

where B is *magnetic field*, n is *proton particle density*, r is *beam radius*, γ_p is *proton relativistic factor*, M is *proton mass*, and r_p is classical *proton radius*, e^2/Mc^2.

For a beam accelerated to *energy* γ_pMc^2 in length L, *synchrotron* loss will balance *energy* gain if

$$L/c = 24\ Mc/r_p{}^2B^2\gamma_p \tag{18}$$

From eqs. (17) and (18) we get:

$$I^3/L = 12 \ (ec)^3 \ (r/L)^2/r_p{}^4 \tag{19}$$

$$P = 2.9 \times 10^{34} \ L^{2/3} \ (\sin q)^{4/3} \ \text{erg/s}, \tag{20}$$

where P is *beam power* and q is *opening angle*. Observations (deRutier, 1990) show that $\theta = 2.6 \times 10^8 \ P^{-0.21}$ so substituting into eq. (20),

$$P = 4.9 \times 10^{35} \ L^{0.52} \ \text{erg/s}. \tag{21}$$

For any *beam* with power larger than that defined by eq (10), the beam will reach a steady state in which it is losing energy to *synchrotron radiation* as fast as it is gaining in by the accelerating field.

[*Synchrotron radiation* (also known as *magnetobremsstrahlung*) is the *electromagnetic radiation* emitted when *relativistic charged particles* are subject to an acceleration perpendicular to their velocity (a ⊥ v). It is produced artificially in some types of particle accelerators or naturally by fast *electrons* moving through *magnetic fields*. The radiation produced in this way has a characteristic *polarization*, and the *frequencies* generated can range over a large portion of the electromagnetic spectrum.

Synchrotron radiation is similar to *bremsstrahlung radiation*, which is emitted by a *charged particle* when the acceleration is parallel to the direction of motion. The general term for radiation emitted by particles in a *magnetic field* is *gyromagnetic radiation*, for which *synchrotron radiation* is the *ultra-relativistic* special case. Radiation emitted by charged particles moving *non-relativistically* in a *magnetic field* is called *cyclotron emission*. For particles in the *mildly relativistic range* (≈ 85% of the speed of light), the *emission* is termed *gyro-synchrotron radiation*.

In *astrophysics*, *synchrotron emission* occurs, for instance, due to *ultra-relativistic motion* of a *charged particle* around a *black hole*. When the source follows a circular geodesic around the *black hole*, the *synchrotron radiation* occurs for orbits close to the photosphere where the motion is in the *ultra-relativistic regime*.]

These beams are unstable to *filamentation*, as is demonstrated in Lerner, 1993[22], since any local increase in *magnetic field* will cause *protons* to lose energy to *synchrotron energy* faster than they are gaining, shrinking their *gyro-radius*, concentrating the *currents* and increasing *magnetic field* further. *A fractal hierarchy of filaments will thus form* with B'^{2}r' and B'γ_p' constants for the whole array, where the primes refer to the *filaments* within the beam.

For most QSO's and AGN's, *emitted power* exceeds, in many cases, greatly exceed the *beam power* P, defined by Eq (21).

$$P = 4.9 \times 10^{35} \, L^{0.52} \text{ erg/s.} \tag{21}$$

For example, for L= 0.3 pc, 10^{18} cm, typical of a QSO, P = 1 x 10^{45}/erg/s, much smaller than the *total energy emission* of *luminous* QSO's (>10^{47} erg/s) while for L= 10^{15} cm, typical of AGN, P = 3 x 10^{43} erg/s, again relatively modest. Thus, *most QSO's and AGNs can be expected to produce arrays of filaments*.

Once such *filamentary beams* leave the accelerating region, the *electrons* and *protons* will lose *energy* rapidly through *synchrotron radiation*. However, such cooling will stop when the *proton synchrotron frequency* is less than the *electron plasma frequency* that is, when the *plasma* becomes opaque to *proton synchrotron radiation*. In a force free *filament*, with all particles moving along *field lines*, the *minimum radius* of the *field lines* is defined by the *proton gyro-radius*, so the *electrons* moving along the same *field lines* radiate at the same *frequency* as the *protons*. Since the *electrons* and *protons* are *counter-streaming*, we have

$$(4\pi n_e e^2/m)^{1/2} = eB\gamma_p'^2 \, \gamma'_e \, /4Mc, \tag{22}$$

and since B'^{2}/n$_e$ = $8\pi\gamma'_p Mc^2$,

$$\gamma'_p{}^5 \, \gamma'_e{}^2 = 8M/m. \tag{23}$$

And, since $\gamma'_e = \gamma'_p$,

$$\gamma'_p = (8M/m)^{1/7} = 3.94. \tag{24}$$

As shown in Lerner (1993)[22], the hierarchical array of *filaments* will expand, maintaining the constancy of B'2r' among the *filaments*, while the value of B'2r' for the array declines. Eventually the expansion of the array ceases when B'2r'= K, the value for the *background plasma*. *Observations* of large-scale *magnetic fields* indicate K ~ 6 x 10^{12} G^2 cm. For example, for a large *galaxy* r = 5 x 10^{22} cm and B = 10^{-5} G, so here K= 5 x 10^{12} G^2. The *number of filaments* were calculated in ref 5 (eqns. 10-11) to also follow a fractal law with fractal dimension D = 2. Thus, within an array the *number* N of *filaments* of *radius* r would be proportional to 1/r and the fraction of sky covered by the *filaments* would be independent of r.

In these filaments, electrons moving along magnetic field lines would be able to absorb and re emit photons at the *electron* synchrotron frequency only by transiently acquiring and losing momentum *perpendicular* to the field lines. *The electrons' energy perpendicular to the lines would be in equilibrium with the background radiation at 2.73 K, but their energy along the field lines would be ~ 2 MeV.* Since the *electron synchrotron frequency* is 1836 times higher than the *proton synchrotron frequency, the highly anisotropic plasma in these filaments can strongly absorb and emit radiation at frequencies thousands of times above the electron plasma frequency.* This behavior is thus extremely different from that in an *isotropic plasma*.

The model predicted (Lerner, 1993[22], eqn. 3) that *a given filament would be highly opaque to frequencies less than the electron synchrotron frequency*, which would be correlated with the *filament radius*:

$$A = \pi h r_e B^2 R f / 6 k T_p M c \gamma_e \gamma_p, \tag{25}$$

where A is the *optical depth* at the *synchrotron frequency*, f = abs (v dR/R dv) = 2, for B^2r = K, and T_p is the *temperature* of *electrons perpendicular to the field lines*. From eqns. (22) and (24)

$$A = 6.6 \times 10^{-12} K. \qquad (26)$$

So, the *opacity* for v < v_s , the *synchrotron frequency*, is independent of B. For K= 6 x 10^{12}, A = 40 and thus *the filaments are highly opaque*. Since

$$v_s = 11 \text{ B MHz}, \qquad (27)$$

the *frequency range* of *opacity* depends mainly on the range of B in the expanded *filament* arrays.

The highest B field in the smallest *filaments* are limited in two ways, as shown in Lerner, 1993[22]. First, the intial γ'_p > 3.94 and second, the *filaments* must survive collisional processes during formation and the expansion of the array. In these *filaments*, the *collision distance* is far greater than the *ion* or *electron gyro-radius*. Because any collision allows particles to move only by one *gyro radius* across the *field lines*, but by one *collision distance* along *field lines, conductivity is enormously greater along field lines than across them*. This leads to the formation of *electrojets*, where *ions* and *counter-streaming electrons* become segregated in separate streams. *Electrojet formation*, driven by the greater rate of collisions that electrons encounter when mixed with ions than when segregated, cannot be countered by electrostatic forces between *ion* and *electron* streams due to the negligible cross-field conductivity. This conductivity decreases with decreasing collision rates, reinforcing the *electron-ion* segregation.

Electrojet formation reduces collisional *energy losses*, but detailed calculations in ref 5 show that the smallest, densest filaments will dissipate before they have time to expand (see eq 15-31, table 1 in Lerner, 1993[22]). These calculations show that the *filament abundance* will start to decline slightly for v_s > 150 GHz and the highest-B filaments will have v_s = 1.4 THz. These *densest filaments* will have

B = 1.2 x 10^5 G and r = 400 cm. Even for these *densest filaments*, lifetime from collisional losses will be very long, > 1400 Gy.

The maximum *filament radius* in the *filament array* will be determined by the *synchrotron radiation decay* of *proton energy* to the stable 4 Gev level. As shown in Lerner, 1992[29], *filaments* with $\gamma'_p > 7$ x 10^6 will have B field too small to decay in < 10 Gy so will not have *electron* that cool down to low T_p. The largest *filaments* will thus have B = 0.07 G and r = 1.3 x 10^{15} cm or 100 AU (astronomical units). These have v_s = 0.8 MHz, but another factor will limit the low-frequency *array opacity*. Radiation with *wavelengths* longer than the minimum *filament radius* of 400 cm will not be scattered by these *smallest filaments* and thus the opacity of the array is expected to start to decline at v < 75 MHz and Lerner, 1992[29], predicts *a low-frequency cut-off of opacity at 150 MHz.*

Comparing these predictions from 30 years ago with current *observations*, we find that on the *high frequency* side, the total cosmic background radiation starts to diverge by about 10^{-4} from a *blackbody* at around 250 GHz and the much hotter, *non-blackbody* IR *background* dominates *total radiation* for n >800 GHz. This is in good agreement with the predicted transition to transparency occurring between 200 and 1400 GHz.

On the *low-frequency* side, the *background radiation* is clearly *non-blackbody* at 80 MHz (Baiesi, *et al.* 2020[30]; Dowell & Taylor, 2018[31]).

[30] Baiesi, M., *et al.* (February, 2020). Possible nonequilibrium imprint in the cosmic background at low frequencies. *Phys. Rev. Research*, 2, 013210; https://doi.org/10.1103/PhysRevResearch.2.013210.

[31] Dowell J., & Taylor, G. B. (May, 2018). The Radio Background below 100 MHz. *Astrophys. J. Lett.*, 858, L9; https://doi.org/10.3847/2041-8213/aabf86.

The data from Wang, L. *et al.*, 2019[24], used in section 3 show that *there is still high opacity at the observed frequency of 150 MHz*, so that transition to transparency occurs between 80 MHz and 150 MHz, in accord with Lerner, 1992[29]. However, the observations are somewhat ambiguous, since at 408 MHz, there is one measurement (Haslam, 1981[32]) of cosmic radiation temperature of 10 ± 3.5 K, which is incompatible with *opacity* at this *frequency*.

[32] Haslam, C. G. T., *et al.* (July, 1981). A 408 MHz all-sky continuum survey. I - Observations at southern declinations and for the North Polar region. *Astronomy and Astrophysics*, 100, 209-19; https://articles. adsabs.harvard.edu/pdf/1981A%26A...100..209H.

But another earlier measurement (Howell & Shakeshaft, 1967[33]) at the same *frequency* yielded a radiation temperature of 3.7 ± 1.2 K, which is compatible with a 2.75 K *blackbody*.

[33] Howell T. F., & Shakeshaft, J. R. (November, 1967). Spectrum of the 3° K Cosmic Microwave Radiation. *Nature*, 216, 753-4; https://doi.org/10.1038/216753a0.

Do known sources of filament arrays— *quasi-stellar object* (QSO)'s and *active galactic nuclei* (AGN)'s and possibly *Herbig-Haro objects*— supply sufficient *current* and *energy* to account for the *observed intergalactic media* (IGM) *opacity* in the *microwave band*? The calculations in Section 3 show that if homogeneity occurs on scales of 1.5 Gpc, then to match *observed opacity* the arrays would have to have an *optical depth* of 1 for a *depth* D of 8×10^{27} cm or 2.6 Gpc. To achieve such an *optical depth* with *filament arrays* with maximum *radii* of r = 1.3×10^{15} cm requires a *fill-factor* of $\pi r/2D = 2.6 \times 10^{-13}$. The *magnetic* and *kinetic energy* in the array is of the order of 6×10^{-5} eV/cm^3, only 0.25% of the *energy* in the *cosmic microwave background* (CMB) and about half that calculated by Lerner for *total cosmic radiation* (CR) from *stars*. The total power from *quasi-stellar object* (QSO)'s and *active galactic nuclei* (AGN)'s has exceeded 10^{41} ergs/s/Mpc or 3.7×10^{-33} erg/sec-cm^3 for the last 10 Gy, which produces an energy density of 7×10^{-4} eV/cm^3. This is a factor of 10 more than in the *filament arrays*. As pointed out in Lerner, 1992[29], the amount of *matter* and *energy* in the hypothesized fractal array of *filaments* is not excessive. With a *mean ion energy* of 4 Gev, even with equipartition of energy between fields and particle, the *total ion density* due to the arrays is 1.5 x 10^{-14}/cm^3, so amounts to < 10^{-7} of *total IGM density*.

Recently, *observations* have provided some direct evidence of the existence of *small, energetic filaments*. In January, 2021, Wang, P., *et al.* (2021)[34] reported observations of a *plasma filament* that are consistent with those predicted 30 years earlier.

[34] Wang, P., *et al.* (June, 2021). Possible observational evidence for cosmic filament spin. *Nat. Astron.,* 5, 839–45; https://doi.org/ 10.1038/s41550-021-01380-6.

They observed five rapid *scintillators* that were aligned in a straight line across a distance of 2 degrees on the sky. The *scintillators*, observed at 945 MHz, had a mean deviation from the straight line of 20", indicating that they lay behind a *plasma filament* that was straight to at least 1 part in 300. By itself, this geometric observation indicated that the *internal magnetic forces* of the *filament* were at least 300 times stronger than the *background magnetic fields* that would tend, over time, to bend the *filament*. Since the *local magnetic field* is about 4 μG, this implies that the *filament internal magnetic field* exceeds 1 mG. Wang, P., *et al.*[34] observed variations in the scintillation over the course of a year that showed the *filament* had a low velocity relative to the Sun close to 10 km/s, which together with other observations led to the conclusion that the *filament* was about 2-4 pc distant. The *radius* of the *filament* at this distance is 0.7-1.5 x 10^{15} cm, just in the upper range of that predicted in Lerner, 1992[29].

Equally significant, Wang, P., *et al.*[34] observed extremely rapid and deep *scintillations* of the sources behind the *filament*, as much as doubling or halving of observed flux over a 15-minute period. Interpreting this as being caused by sub-filaments moving across the source, the observed 10 km/s velocity implies objects with *radii* of the order of only 10^9 cm. Since the rapid oscillations are continuous and exceed 50% of signal strength, the number of *sub-filaments* in the overall *filament* must be on the order of the ratio of the *radii*: ~ 10^{15} cm/10^9 cm = 10^6, as predicted for such arrays in Lerner, 1992[29]. A significantly smaller number of *filaments* would lead to intermittent rather than continuous oscillations, while a significantly larger number will smear the *scintillations* to produce smaller amplitudes than those observed.

For the *observing frequency* of 945 MHz, eq. (27) combined with K = B^2r = 6 x 10^{12}, predicts a filament radius r of 0.8 x 10^9 cm, in good agreement with these recent *observations*. Significantly, other recent *observations* (Osterloo, *et al.*, 2020[35]) have confirmed the existence of the predicted *small-radius filaments* on the order of 10^9 cm.

[35] Osterloo, T. A., *et al.* (September, 2020). Planck 2018 results. VI. Cosmological parameters. *Astronomy and Astrophysics*, 641, L4, A6; https://doi.org/10.1051/0004-6361/201833910.

Observations of *scintillation* (Koay, *et al.*, 2019)[36] have extended up to *higher frequencies* of 15 GHz, *showing that plasma filaments exist that can affect radiation in the range where the cosmic microwave background (CMB) is closest to a blackbody.*

[36] Koay, J. Y., *et al.* (November, 2019). The presence of interstellar scintillation in the 15 GHz interday variability of 1158 OVRO-monitored blazars. *MNRAS*, 489, 4, 5365-80; https://doi.org/10.1093/mnras/stz2488.

Based on the small distance to this *filament*, Wang, P., *et al.*[34] estimate that the *optical depth* generated by such *filaments* is ~ 10^{-3}/pc, which means that they will cover about 0.4 of the sky in the direction *perpendicular to the galactic plane*, assuming they are confined to within 400 pc of the plane. If we hypothesize such *filaments* exist in the *intergalactic media* (IGM) in numbers proportional to *ion density*, with the local *ion density* being about 0.1/cm^3, then the *optical depth* in the IGM at an *ion density* of 6 x 10^{-8}/cm^3 would 0.6/Gpc. This would be 30% less than the *optical depth* estimated above from extrapolating actual *observations* of RF *absorption* in the IGM, well within the uncertainties in both estimates.

These observations can be explained by *small-angle deflections* by the *plasma filaments*, and that is indeed the explanation assumed in the papers reporting the *observations*. However, they are also entirely consistent with the *large-angle scattering*, actually *absorption* and *re-emission*, hypothesized by Lerner, 1992[29], 1993[22]. Further observations and modeling will be required to unequivocally distinguish between these two possibilities. But if the *filaments* are generating *large angle scattering*, they would be entirely consistent with both the phenomena responsible for the observed IGM absorption and with the *Galactic Origin of Light Elements* (GOLE) hypothesis *plasma filament model*.

8.4 Conclusions on the CMB

The observation of radio-frequency absorption shows that the production of an isotropic black-body cosmic microwave background (CMB) *occurs in the present-day universe, although it does not preclude an earlier isotropization by a Big Bang. However, these observations do cast strong doubt on the hypothesis that the observed CMB fluctuations are created by a Big Bang. The fluctuations are instead dominated by fluctuations in density of the absorbing medium in the intergalactic media* (IGM). It is significant to note that even the *anomalous microwave emission* (AME) already observed in the Milky Way (MW), if typical of other galaxies, would by itself produce *fluctuations* on the level of those observed. The AME is strongly correlated with H column density at a level of 3×10^{-18} Jy cm^2. With typical IGM column density of the order of 3×10^{21}/cm^2, AME levels from *galactic clouds* like those in the MW would be expected to be ~10 kJy, while *fluctuations* in the *cosmic microwave background* (CMB) are ~ 4kJy. The actual *fluctuations* produced by *high-opacity* objects would of course depend on their nature and distribution, and cannot be predicted without further detailed calculations.

However, *the evidence presented in this paper shows that only about 8 ± 4% of RF radiation from > 600 Mpc distance reaches Earth*. Such a strong absorption (or high-angle scattering) has the potential to reduce to low levels any trace of the high degree of inhomogeneity in the IGM, just as inhomogeneities in cloud density become invisible in a heavy fog. However, whether the actual observed inhomogeneities in galaxy distribution can reproduce the observed inhomogeneities in the CMB requires considerable additional research.

So, in contrast to the *Big Bang hypothesis* (BBH) model of the *cosmic microwave background* (CMB), the *Galactic Origin of Light Elements* (GOLE) hypothesis model, *in which the energy for the CMB is supplied by the same fusion reactions that produce the observed heli*um, and the energy is isotropized and thermalized by the *intergalactic media* (IGM), *produces no contradictions with observations*, and is confirmed by evidence for the opacity of the IGM at CMB frequencies.

Of course, considerable further work is required to make more detailed comparisons with observations of the fluctuations expected from an isotropizing medium in the present-day universe, and to compare more detailed models of *filament arrays* with the new observations that will emerge from surveillance with the new phased array instruments. But research resources would be better directed to these questions than to the *Big Bang hypothesis* (BBH) model of the *cosmic microwave background* (CMB), which is now contradicted by *observations* for all quantitative predictions.

9. Discussion

The updated reassessment presented here, comparing the *galactic origin of light elements* (GOLE) *hypothesis* with the *Big Bang hypothesis* (BBH), confirms that the GOLE hypothesis is able to correctly predict the abundances of He, Li, D, C, and O, as well as the correlation of Be with O, while BBH only correctly predicts D and is completely contradicted by He and Li data. Specifically, it has been demonstrated in section 4 that *ultraluminous infrared galaxies* (ULIRGs) data confirm the hypothesis that *forming galaxies* produce He at the high rate predicted by GOLE, and C and O in amounts in accord with GOLE predictions. Updating the GOLE model based on the latest theoretical calculations and observational data confirms that the model predicts the observed abundances of He, D and Li. Taking into account extensive observations showing that most CR are directed downwards towards their star, neutrino data have provided evidence of the amount of CR hypothesized by GOLE. Independent confirmation of this CR production rate is obtained from the observed correlation of O and Be abundance.

The other direct tests of the *existence or non-existence of a hot, dense phase* of cosmic evolution have similarly been demonstrated to be consistent with GOLE. The *antimatter/baryon non conservation problem* is eliminated without BBH, as pointed out in section 5. Combined with the *hypothesis of energy loss of EM radiation*, non-BB evolution also correctly predicts the *constancy of surface brightness*, and this constancy has now been confirmed by the entirely independent analysis of Whitney, *et al.* (2020)[37].

[37] Whitney, A., *et al.* (October, 2020). Surface Brightness Evolution of Galaxies in the CANDELS GOODS Fields up to $z \sim 6$: High-z Galaxies Are Unique or Remain Undetected. *Astrophys. J.*, 903, 14; https://doi.org/10.3847/1538-4357/abb824.

In section 6 the *SN1a luminosity- distance relationship* predicted by the *non-expansion hypothesis* is shown to be closer to the *SN1a data* once the newly-discovered correlation of SN properties with galactic age is taken into account. In section 7 for the first time, it is demonstrated that, *when the galactic age dependence is taken into account, SN1a* observed widths are constant with z, in accord with *non-expansion* predictions and in *contradiction with* BBH. *In all of these additional direct tests, predictions based on BBH are contradicted by the data, and those based on no hot, dense phase of evolution (which we refer to as GOLE) are confirmed by the data.*

Other data sets have been cited in support of the BBH, specifically observations of the *cosmic microwave background* (CMB). It is widely argued that the *energy density, isotropy* and *black body spectrum* of the CMB can't be explained by anything other than the BBH.

However, in section 4.2 we show that the *energy* generated in the production of the observed amount of He is comparable to, and might even exceed, the observed *energy* in the CMB, so this *energy density* does not require the BBH. The *isotropy* and *black-body spectrum* of the CMB are evidence *not specifically for a hot dense stage of cosmic evolution*, but are instead evidence that *the universe at some point in its history had a high optical depth in the microwave band*. There is convincing evidence first published in 1990 (Lerner 1990[21], 1993[22]) and elaborated in Lerner, 2021, that the present universe has this high optical depth in this wavelength range, so *no past high-density epoch* is required.

Specifically, these papers show that the *luminosity* in the *radio waveband* of *galaxies* with a given *infrared luminosity* falls rapidly with increasing *distance* from earth. This fall occurs with 600 Mpc of earth, in a distance that is impossible to explain by any evolutionary change. *This observational evidence can only be explained by an absorption or scattering of radio-band radiation in the IGM which can account for a high optical depth in the microwave band in the present-day universe*, eliminating the need for a high-density phase in the past. Lerner, 1992[29], describes in detail *a plasma mechanism for the absorption and re-emission of microwave-band radiation by magnetized filaments in the intergalactic media (IGM)*. Thus, known processes occurring in the present-day universe can account for the main features of the *cosmic microwave background* (CMB).

At the same time, the quantitative predictions of the BBH model of the CMB *have been increasingly contradicted by observation*. The CMB fluctuations are not Gaussian on the largest scales as BBH requires (Schwarz, 2016)[3]; the largest-scale modes are a poor fit to BBH predictions (Schwarz, 2016)[3], and the CMB fluctuations are not best fit by the *flat universe* required by BBH inflation (Handley, 2019[8]; Park & Ratra, 2019[9]; DiValentino, *et al*, 2020[10]). Finally, as has been widely noted, *the BBH predictions of the Hubble relation, based on the CMB, are wrong* (Riess, 2020)[11], as are *the best-fit predictions of matter density* (DiValentino, *et al.*, 2020[10]). Thus, not only is the BBH not required to explain the *cosmic microwave background* (CMB), *its specific quantitative predictions are at present in contradiction with observation.*

In assessing whether the BBH or GOLE is valid it is important to note that observations over, especially, the last five years demonstrated a *growing number of additional contradictions between other BBH predictions and observations*, specifically involving the *"age problem"* and *structure formation* with *dark, non-baryonic, matter* (DM). The *age problem* is the existence of objects in the universe that are older, in some cases far older, than the time since the hypothesized *Big Bang*.

In addition, BBH structure formation requires the existence of *dark, non-baryonic, matter* (DM). This DM hypothesis itself has encountered growing contradictions with observations including no lab evidence of DM particles (Adhikari, et al., 2018[13]; Aprile et al. (XENON Collaboration), 2018[14]; Liu, Chen, & Ji, 2017)[38]; evidence against DM dynamic viscosity (Oehm & Kroupa, 2018)[39]; disk alignments of galaxy satellites (Santos-Santos, Domínguez-Tenreiro, & Pawlowski, *et al.*, 2019[40]; Müller, *et al.*, 2018[41]); and no DM in small galaxies (Mancera Pina, *et al.*, 2019[42]).

[38] Liu, J., Chen, X., & Ji, X. (2017). Current status of direct dark matter detection experiments. *Nature Physics*, 13, 212-6; https://doi.org/ 10.1038/nphys4039.

[39] Oehm, W., & Kroupa, P. (November, 2018). Constraints on the existence of dark matter haloes by the M81 group and the Hickson compact groups of galaxies. arXiv:1811.03095; https://doi.org/ 10.48550/arXiv.1811.03095.

[40] Santos-Santos, I., Domínguez-Tenreiro, R., *et a*l. (April, 2020). Planes of Satellites around Simulated Disk Galaxies. I. Finding High-quality Planar Configurations from Positional Information and Their Comparison to MW/M31 Data. *Astrophys. J.*, 897, 1; https://doi.org/10.3847/1538-4357/ab7f29.

[41] Müller, *et al*., (February, 2018). A whirling plane of satellite galaxies around Centaurus A challenges cold dark matter cosmology. *Science*, 359, 6375, 534-7; https://doi.org/10.1126/science.aao1858.

[42] Mancera Piña, P. E., *et al*., (September, 2019). Off the baryonic Tully-Fisher relation: a population of baryon-dominated ultra-diffuse galaxies. *Astrophys. J. Lett.*, 883, L33; https://doi.org/10.3847/2041-8213/ab40c7.

Such age problems do not exist for the GOLE hypothesis, which eliminates the hypothesis of an initial dense, hot epoch for the universe. By including the well-known effect of magnetic fields and electric currents, and by discarding the BBH, including DM, the main features of the formation of cosmic structure have been accurately predicted (Alfven, 1978[43], 1981[44]; Lerner, 1986[45]; 2022).

[43] Fälthammar, C. G., Akasofu, S. I. & Alfvén, H. (1978). The significance of magnetospheric research for progress in astrophysics. *Nature*, 275, 185–8; https://doi.org/10.1038/275185a0.

[44] Alfvén, H. (1981). *Cosmic Plasma*. Springer, Dordrecht.

[45] Lerner, E. J. (December, 1986). Magnetic Vortex Filaments, Universal Invariants and the Fundamental Constants. *IEEE Trans. Plasma Sci.*, Special Issue on Cosmic Plasma, PS-14, 6, 690-702; https://doi.org/10.1109/TPS.1986.4316620.

The application of this model to the early stage of galaxy formation leads to the GOLE hypothesis (Lerner, 1988[19], 1989[46]).

[46] Lerner, E. J. (April, 1989). Galactic Model of Element Formation. *IEEE Transactions on Plasma Science*, 17, 3, 259-63; https://doi.org/10.1109/27.24633.

At the present time, no predictions of the BBH are in quantitative agreement with subsequently published observational date, except the D abundance predictions. There are at least 16 independent observational contradictions to these predictions (Table 2). This extremely poor performance of BBH exists despite the fact that the BBH currently requires at least four hypotheses not based on laboratory-validated physics: the inflation field, non-baryonic matter (DM), dark energy and baryonic non-conservation. The GOLE hypothesis, in contrast, requires only one new hypothesis—that EM radiation loses energy as it travels long distances.

This situation has developed mainly over the last seven years. Before about 2015, it could be argued on the basis of peer-reviewed publications that BBH had more verified predictions than observational contradictions, that contradictions were isolated "anomalies" in a generally well supported theory. In Fig 16, we plot the number of independent predictions of BBH that were reported in peer-reviewed papers as verified by observations made after the predictions, against the year (long-dashed line). Similarly, we plot the number of BBH predictions that were contradicted by observations, as reported in peer-reviewed journals (short-dashed line). When new data contradicts predictions previously reported as confirmed, the total number of confirmed predictions decreases accordingly. The data is the same as that in Table 2. By 2006 there were 10 independent confirmations of BBH predictions reported in the literature and only two clear contradictions, to the Li and He data. We also plot the number of confirmations of GOLE predictions in peer-reviewed journals (solid line) based on the data in Table 3. As of 2006, there were four independent predictions of GOLE reported as confirmed by subsequent observations.

A number of researchers, including the present author (Lerner, 1993[22]), had published critiques of the validity of many of these claimed confirmations, including questioning the methodology of introducing ad-hoc concepts such as dark energy. However, the number of reported confirmed BBH predictions remained well above the number of clear-cut contradictions in the first decade of the century.

In 2014, this situation began to change at an accelerating pace. Analyses of surface brightness data showed unequivocal contradiction with BBH predictions (Lerner, Falomo, Scarpa 2014[47]) and the release of Planck data confirmed many contradictions with the *cosmic microwave background* (CMB) predictions of BBH (Schwarz, 2016[3]).

[47] Lerner, E. J., Falomo, R., & Scarpa, R. (2014). UV surface brightness of galaxies from the local universe to $z \sim 5$. *International Journal of Modern Physics* D, 23, 06, 1450058; https://doi.org/10.1142/S0218271814500588; see above.

Deeper surveys revealed large scale structures incompatible with BBH age predictions (Shirokov, *et al.*, 2016[27]). Multiple contradictions with DM predictions were reported, as noted above. Finally, in the 2019-2020, widely-publicized contradictions with BBH predictions of the Hubble constant, based on the CMB, led to widespread acknowledgment of a "crisis in cosmology" (DiValentino, *et al.*, 2020[10]). In the present paper, the SNIa widths are added to the list of contradictions. During the same period, three independent predictions of the GOLE theory were confirmed, and there have continued to be no published contradictions of any GOLE predictions, as documented in the present papers.

At present, the focus of researchers' attention has remained overwhelmingly on the contradictions with BBH predictions arising from the CMB data. Both the long-standing contradictions with Li and He predictions and the more recent contradictions with surface brightness, large scale structure and DM have been largely ignored. This is to a certain extent a product of the specialization of the field, in which many researchers are simply unaware of developments outside their own area of research. The present paper and the author's recent paper on structure formation (Lerner, 2022) constitute a first attempt to overcome this problem and bring together the new results reported here

317

with previously-published results to show the true situation of the BBH and GOLE alternatives.

Given that the accumulation of contradictions of BBH predictions in all of these data sets over the past eight years, the obvious question is why the BBH remains the accepted model, used by the vast majority of cosmology researchers. It should be clear from the summaries in this paper and the extensive literature that they rest on, that the answer does not lie in the scientific validity of the hypothesis, which has clearly been falsified by observation. It lies instead in the sociology of science.

As Merritt, (Merritt, 2017)[48], among others, has pointed out, the field of cosmology has for a long time taken a "conventionalist" approach.

> [48] Merritt, D. (2017) *Cosmology and Convention*. Studies in History and Philosophy of Science Part B: *Studies in History and Philosophy of Modern Physics*, 57, 41-52; https://doi.org/10.1016/j.shpsb. 2016.12.002.

Instead of seeking to validate or falsify scientific hypotheses, the conventionalist approach takes certain paradigms as given or "orthodox" and defends them against multiple falsifications by observation. It does this with methods such as ad hoc hypotheses, treating contradictions as isolated "anomalies" or by simply ignoring them.

The standard answer to each of the severe contradictions enumerated in table 2 is "BBH is overwhelmingly confirmed by many other data sets, so this one is just an anomaly that has not yet been explained." Such a gross violation of the scientific method of course blocks progress in this field, as occurred for centuries with the Ptolemaic cosmology.

To overcome this widespread, but unscientific response, *it has been essential in the present paper to comprehensively present in a single place a full comparison with all data sets of the BBH and the alternative—the abandonment of the BBH.* This admittedly results in a lengthy paper, but is unavoidable in the present state of cosmological discussion.

It is also the reason why this paper must be published on Arxiv, rather than in a peer-reviewed publication. It is against the policy of leading peer-reviewed publications to publish such a paper. An earlier version of this present paper was rejected by *MNRAS*, for example, after a long review process, in which I satisfied all of the reviewer's major objections, (and all but one of the minor objections). The anonymous senior editor rejected the paper, writing that "There are many journals which would be interested in publishing a well-argued synthesis of existing evidence against the standard hot big bang interpretation. But MNRAS, with its focus on publication of significant new astronomical results, is not one of them." I confirmed with Dr. Flowers, the editor-in-chief, that this is indeed MNRAS's policy.

The reaction of researchers to the growing contradictions with observation that has become widely acknowledged as the "crisis in cosmology" also reflects a conventionalist approach. Instead of viewing the multiple contradictions of predictions by observation as falsifications of the basic BBH, paper after paper proposes additional ad-hoc assumptions such as early dark energy, new early dark energy, phenomenologically emergent dark energy, interacting dark energy, fuzzy dark matter, decaying dark matter dark forces and so on. This drives cosmology into a dead-end.

It is important to note that the adoption of any of the elaborations on dark energy, even if they were to conform to all existing data, would entirely destroy the predictive power of the BBH. While previous additions to the BBH have added one free parameter at a time, the adoption of new concepts of dark energy would be chosen from among an infinite set of possible functions. Since any predictive theory has to have fewer free parameters than independent observations, the adoption of a theory with an infinite set of free parameters would eliminate BBH as a predictive and thus scientific theory.

But the overwhelming adherence to this conventionalist approach raises the deeper question— *why has cosmology, in particular among scientific fields, become paralyzed for decades by this conventionalist straightjacket?* In other fields of science, dearly-held and entrenched paradigms have been overthrown, but not in cosmology, despite decades of deepening observational contradictions, and the existence of an alternative hypothesis—namely, the simple rejection of the Big Bang hypothesis.

It is beyond the scope of this article to discuss in full the reasons for this rigidity, which has been explained in detail elsewhere (Alfven, 1978[43]; Lerner, 1992[29]; López Corredoira & Perelman, 2008[49]).

[49] López Corredoira, M., & Perelman, C. C. (Eds.) (2008). *Against the Tide. A Critical Review by Scientists of How Physics and Astronomy Get Done.* Universal Publishers, Boca Raton, Florida, USA; https://philsci-archive.pitt.edu/4046/1/againsttide.pdf.

However, two key points should be briefly outlined. First, in almost all fields of science, the ultimate test of validity is in the application to technology. The validity of Maxwell's equations is demonstrated each time a light switch is thrown, those of quantum mechanics every time you look at a computer screen. The effectiveness of new paradigms in technology ultimately drives acceptance.

Cosmology in the past half century and more has appeared to be effectively isolated from applications to technology. While discoveries in astrophysics in the past had obvious implications for technology, such as the discovery in the 1930's of nuclear fusion as the source of energy of the stars, no such links are widely recognized today. The real links that do exist between *plasmas* in the cosmos and those on earth are unknown to all but a few researchers in cosmology. So, the possibility of paradigm change as demonstrated by technological advances has been absent.

Second, and even more important, is the concentration of the funding for cosmological research in a very small number of sources. In other fields of science, a variety of funding sources, including many different governmental sources, allow for a diversity of models. But in cosmology, largely because of its perceived lack of application, such numerous different funding sources don't exist.

Funds for astronomical research and time on astronomical satellites are allocated almost exclusively by very few governmental bodies, such as the *National Science Foundation* (NSF) and *National Aeronautics and Space Administration* (NASA) in the United States. The review committees that allocate these funds are tightly controlled by adherents of the *Big Bang* theory, who refuse to fund anything that calls their life's work into question. It is no secret that, today, anyone who questions the *Big Bang*, who develops alternatives to the *Big Bang*, or, for the most part, who even investigates evidence that contradicts the *Big Bang*, will not receive funding.

This concentration of funding sources creates a strong *"Emperor's-New-Clothes"* effect where those who don't see the beauty of the BBH are deemed incompetent and thus unworthy of funding. It should be remembered that it was only the Emperor's unquestioned power over people's livelihoods that led to the dominant theory that the clothes were beautiful.

In Anderson's immortal fable, it is ridicule that finally undoes the Emperor. Given the widespread attention to the Crisis in Cosmology, perhaps researchers will soon see the inflationary ΛCDM *Big Bang* (with its contradictions to 16 different sets of data) as equally, or more, ridiculous than the fictional Emperor. Perhaps at that point the funding of cosmology can be reformed to be more diverse and less dominated by conflicts of interest. Certainly, having funding requests reviewed by astrophysicists from outside the field of cosmology would be a first step the right direction.

The only route out of the crisis is to recognize that BBH has in recent years been falsified. This route would then involve elaborating and testing against observations a cosmological model without a *Big Bang*. A number of tasks on this route are clear right now. Uncertainties in the theory of light elements production can be reduced through observations of stars and planetary nebulae, as well of as young galaxies (*ultraluminous infrared galaxies* (ULIRGs)). Modeling of stellar evolution of initially pure H stars is also needed. Observational tests can be devised to better constrain the production of cosmic rays by stars of different masses. For the *cosmic microwave background* (CMB), extensive modeling, with significant resources, is needed to compare with observations the power spectrum and other features of a predicted CMB produced by thermalized stellar radiation and scattered by the IGM. As pointed out in section 6, space-based experiments are possible to directly test the hypothesis that EM radiation loses energy with distance. On this route, the crisis in cosmology will be resolved on the basis of known physical laws and observations, paving the way to a model that makes increasingly useful predictions of subsequent observations.

Table 3. Predictions of GOLE

Data set/Prediction	Year confirmation
Large Scale Structure-nr relation	1989
Radio-IR absorption	1990
Li abundance	2003
D abundance	2003
He abundance	2007
surface brightness constancy	2014
SNIa luminosity-distance	2021
ULIRG luminosity	2021

Table 3. Key quantitative predictions of the GOLE. The second column shows the year in which peer-reviewed journals first published evidence in papers by the present author that the predictions were confirmed by data published after the predictions were made.

10. Conclusions

A re-analysis of data and predictions of the *galactic origin of light elements* (GOLE) *hypothesis* almost 40 years after it was first proposed confirms that *it has accurately predicted light element abundances*. Its predictions, *assuming the nonexistence of a hot, dense period of cosmic evolution*, have as well been confirmed by many other data sets. *Only a single new assumption of EM energy loss with distance is required.*

> [This is provided by *Zwicky's tired-light theory as elaborated by Compton*. See Underwood, T. G. (November, 2024). *Cosmological Redshift of Light*.]

No incompatibilities with data have been found.

In contrast, almost 60 years after the *Big Bang hypothesis* was proposed for the origin of light elements, *observations have increasingly diverged from its predictions for He and Li abundances*, leaving only D as a correct prediction. The *Big Bang hypothesis'* other predictions have been *abundantly contradicted by at least 16 data sets*, with divergence between prediction and observation increasing greatly over the past several years. Other than D abundance, the Big Bang predictions are not in accord with any data sets. Observations, taken as a whole, clearly exclude the hypothesis of a hot, dense (*Big Bang*) phase of universal evolution. The only solution to the crisis in cosmology is to recognize that *the Big Bang never happened*.

www.ingramcontent.com/pod-product-compliance
Lightning Source LLC
Chambersburg PA
CBHW050527190326
41458CB00045B/6729/J